Agile Machine Learning

Effective Machine Learning Inspired by the Agile Manifesto

Eric Carter

Matthew Hurst

Apress®

Agile Machine Learning: Effective Machine Learning Inspired by the Agile Manifesto

Eric Carter
Kirkland, WA, USA

Matthew Hurst
Seattle, WA, USA

ISBN-13 (pbk): 978-1-4842-5106-5
https://doi.org/10.1007/978-1-4842-5107-2

ISBN-13 (electronic): 978-1-4842-5107-2

Managing Director, Apress Media LLC: Welmoed Spahr
Acquisitions Editor: Joan Murray
Development Editor: Laura Berendson
Coordinating Editor: Jill Balzano

Distributed to the book trade worldwide by Springer Science+Business Media New York, 233 Spring Street, 6th Floor, New York, NY 10013. Phone 1-800-SPRINGER, fax (201) 348-4505, e-mail orders-ny@springer-sbm.com, or visit www.springeronline.com. Apress Media, LLC is a California LLC and the sole member (owner) is Springer Science + Business Media Finance Inc (SSBM Finance Inc). SSBM Finance Inc is a **Delaware** corporation.

For information on translations, please e-mail rights@apress.com, or visit http://www.apress.com/rights-permissions.

Apress titles may be purchased in bulk for academic, corporate, or promotional use. eBook versions and licenses are also available for most titles. For more information, reference our Print and eBook Bulk Sales web page at http://www.apress.com/bulk-sales.

Any source code or other supplementary material referenced by the author in this book is available to readers on GitHub via the book's product page, located at www.apress.com/9781484251065. For more detailed information, please visit http://www.apress.com/source-code.

Printed on acid-free paper

Matthew would like to dedicate this book
to his family

—Wakako and Hana.

Eric would like to dedicate this book to his wife Tamsyn.

Table of Contents

About the Authors

 **Eric Carter** has worked as a Partner Group Engineering Manager on the Bing and Cortana teams at Microsoft. In these roles he worked on search features around products and reviews, business listings, email, and calendar. He currently works on the Microsoft Whiteboard product.

Matthew Hurst is a Principal Engineering Manager and Applied Scientist currently working in the Machine Teaching group at Microsoft. He has worked in a number of teams in Microsoft including Bing Document Understanding, Local Search and in various innovation teams.

About the Technical Reviewer

James McCaffrey works for Microsoft Research in Redmond, Washington. James has a PhD in computational statistics and cognitive psychology from the University of Southern California, a BA in experimental psychology from the University of California at Irvine, a BA in applied mathematics, and an MS in computer science. James is also the Senior Technical Editor for Microsoft MSDN Magazine, one of the most widely-read technical journals in the world. James learned to speak to the public when he worked at Disneyland during his college days, and he can still recite the Jungle Cruise ride narration from memory. His personal blog site is https://jamesmccaffrey.wordpress.com.

Introduction

This book was born out of a fortuitous meeting. In July of 2012, Eric Carter had just returned to the U.S. following a three-year assignment in Germany launching a shopping search product for Microsoft to the European market. He was sorely disappointed because the effort he had led in Europe was shutting down and so began looking for a new gig. While exploring opportunities in Bing, Microsoft's search engine, he met Matthew Hurst. Matthew had joined Microsoft as a member of Live Labs, an innovation group tasked with exploring novel solutions and applications around search, the cloud and connected technologies. From there he'd worked on various incarnations of maps and local search, often on features connecting text with location. What followed was a complementary partnership that vastly improved the quality of Bing's local search and ultimately led both on a learning journey of how data engineering projects benefit from the application of *Agile principles*.

The Agile Manifesto (or, *The Manifesto for Agile Software Development*, to give it its full title) came into being in 2001 as a collaboration of the seventeen signatories[1]. It is summarized as four values (*individuals and interactions* are valued over process and tools, *working software* over comprehensive documentation, *customer collaboration* over contract negotiations, and *responding to change* over following a plan) and twelve principles. In this book, we examine each of the principles in turn, and relate them to our experiences in working with data and inference methods in a number of projects and contexts.

When the authors met, Bing's local search product was very much a work-in-progress. The quality of the catalog of local businesses was improving, but it was still far behind the Google, market leader at the time. Matthew was on the local search data team and he, along with other members of the team, had been exploring some innovative ideas to better leverage the web and integrate machine learning to dramatically improve the catalog. Eric saw a number of compelling challenges in the local search space as it existed in Bing, and decided to join as the engineering manager of Bing's local data team.

---

[1] Kent Beck, Mike Beedle, Arie van Bennekum, Alistair Cockburn, Ward Cunningham, Martin Fowler, James Grenning, Jim Highsmith, Andrew Hunt, Ron Jeffries, Jon Kern, Brian Marick, Robert C. Martin, Steve Mellor, Ken Schwaber, Jeff Sutherland, and Dave Thomas.

At this point in his career, Eric was no stranger to managing teams at Microsoft, having been part of several Visual Studio related products and the now dismantled "shopping search" project. However, it was during his time working with Visual Studio that he discovered the intrinsic value of Agile, and how much more efficient and happy teams were when following Agile principles. Wanting to bring that to his new team, he found himself in a quandary— how do you apply Agile to a team that is more about producing data than producing software? What would it take to bring Agile to a data engineering team?

It wasn't easy. At first, Agile seemed like an invading foreign agent. The team culture was about big ideas discovered through experimentation, long horizon research, and a lot of trial and error science projects—all seemingly contrarian to Agile principles like scrum, iterative development, predictability, simplicity, and delivering working software frequently. With a team focused on producing an exceedingly accurate database of all the businesses in the world, defining "done" was nothing short of impossible. After all, the singular constant in data is that it contains errors—the work is literally never done. Faced with challenges such as communicating to stakeholders how and where the team was making progress, determining whether a particular development investment was worth making, and ensuring that improvements are delivered at a regular but sustainable pace, it became apparent that a modern Agile approach was critical. But how does one apply Agile in a team comprised of data scientists and traditional engineers all working on data-oriented deliverable?

Traditional Agile processes were intended to reduce unknowns and answer questions such as "what does the customer want" and "how can software be delivered reliably and continuously". But in this new project, new world, we already knew what the customer wanted (a perfect catalog of local businesses[2]) but we needed to answer questions such as "What's in the data?" and "What are we capable of delivering based on that data?" We needed Agile approaches, but revised for a modern, mixed talent, data engineering team.

As we navigated through "next-generation" machine learning challenges, we discovered that, without question, Agile principles can be applied to solve problems and reduce uncertainty about the data, making for a much happier and efficient team.

---

[2]As we will later see in Chapter 3, what the customer wanted wasn't quite as simple as I thought.

Our hope in bringing together this modernized version of Agile methodologies is that the proven guidance and hard earned insights found in this book will help individuals, technical leads and managers be more productive in the exciting work that is happening in machine learning and big data today.

June 2019

Eric Carter

Matthew Hurst

CHAPTER 1

Early Delivery

*Our highest priority is to **satisfy the customer** through **early** and **continuous** delivery of **valuable [data]**.*

—agilemanifesto.org/principles

Data projects, unlike traditional software engineering projects, are almost entirely governed by a resource fraught with unknown patterns, distributions, and biases – the data. To successfully execute a project that delivers value through inference over data sets, a novel set of skills, processes, and best practices need to be adopted. In this chapter, we look at the initial stages of a project and how we can make meaningful progress through these unknowns while engaging the customer and continuously improving our understanding of the data, its value, and the implications it holds for system design.

To get started, let's take a look at a scenario that introduces the early stages of a project that involved mining local business data from the Web which comes from the authors' experience working on Microsoft's Bing search engine. There are millions of local business locations in the United States. Approximately 50%[1] of these maintain some form of web site whether in the form of a simple, one-page design hosted by a drag-and-drop web hosting service or a sophisticated multi-brand site developed and maintained by a dedicated web team. The majority of these businesses update their sites before any other representation of the business data, driven by an economic incentive to ensure that their customers can find authoritative information about them.[2] For example, if their phone number is incorrect, then potential customers will not be able to reach them; if they move and their address is not updated, then they risk losing existing clients; if their business hours change with the seasons, then customers may turn away.

---

[1]Based on analysis of local business feed data.

[2]A survey of businesses showed that about 70% updated their web sites first and then other channels such as social media or Search Engine Optimization (SEO) representatives.

E. Carter and M. Hurst, *Agile Machine Learning*, https://doi.org/10.1007/978-1-4842-5107-2_1

A local search engine is only as good as its data. Inaccurate or missing data cannot be improved by a pretty interface. Our team wanted to go to the source of the data – the business web site – to get direct access to the authority on the business. As an aggregator, we wanted our data to be as good as the data the business itself presented on the Web. In addition, we wanted a machine-oriented strategy that could compete with the high-scale, crowd-sourced methods that competitors benefitted from. Our vision was to build an entity extraction system that could ingest web sites and produce structured information describing the businesses presented on the sites.

Extraction projects like this require a schema and some notion of quality to deliver a viable product,[3] both determined by the customer – that is, the main consumer of the data. Our goal was to provide additional data to an existing system which already ingested several feeds of local data, combining them to produce a final conflated output. With an existing product in place, the schema was predetermined, and the quality of the current data was a natural lower bound on the required quality. The schema included core attributes: business name, address, phone number, as well as extended attributes including business hours, latitude and longitude, menu (for restaurants), and so on. Quality was determined in terms of errors in these fields.

The Metric Is the Customer   The first big shift in going from traditional agile software projects to data projects is that much of the role of the customer is shifted to the metric measured for the data. The customer, or product owner, certainly sets things rolling and works in collaboration with the development team to establish and agree to the evaluation process. The evaluation process acts as an oracle for the development team to guide investments and demonstrate progress.

Just as the customer-facing metric is used to guide the project and communicate progress, any component being developed by the team can be driven by metrics. Establishing internal metrics provides an efficient way for teams to iterate on the

---

[3]Be extremely wary of projects that haven't, won't, or can't define a desired output. If you find yourself in the vicinity of such a project, run away – or at least make it the first priority to determine exactly what the project is supposed to produce.

inner loop4 (generally not observed by the customer). An inner metric will guide some area of work that is intended to contribute to progress of the customer-facing metric.

A metric requires a data set in an agreed-upon schema derived from a sampling process over the target input population, an evaluation function (that takes data instances and produces some form of score), and an aggregation function (taking all of the instance results and producing some overall score). Each of these components is discussed and agreed upon by the stakeholders. Note that you will want to distinguish metrics of quality (i.e., how correct is the data) from metrics of impact or value (i.e., what is the benefit to the product that is using the data). You can produce plenty of high-quality data, but if it is not in some way an improvement on an existing approach, then it may not have any actual impact.

Getting Started

We began as a small team of two (armed with some solid data engineering skills) with one simple goal – to drive out as many unknowns and assumptions as possible in the shortest amount of time. To this end, we maximized the use of existing components to get data flowing end to end as quickly as possible.

Inference   We use the term "inference" to describe any sort of data transformation that goes beyond simple data manipulation and requires some form of conceptual modeling and reasoning, including the following techniques:

- Classification: Determining into which bucket to place a piece of data

- Extraction: Recognizing and normalizing information present in a document

- Regression: Predicting a scalar value from a set of inputs

- Logical reasoning: Deriving new information based on existing data rules

---

[4]An "inner loop" is the high-frequency cycle of work that developers carry out when iterating on a task, bringing it to completion. It is a metaphorical reference to the innermost loop in a block of iterative code.

Structural transformations of data (e.g., joining tables in a database) are not included as inference, though they may be necessary components of the systems that we describe.

Adopting a strategy of *finding* technology rather than *inventing* technology allowed us to build something in a matter of days that would quickly determine the potential of the approach, as well as identify where critical investments were needed. We quickly learned that studying the design of an existing system is a valuable investment in learning how to build the next version.

But, before writing a single line of code, we needed to look at the data. Reviewing uniform sample of web sites associated with businesses, we discovered the following:

- Most business web sites are small, with ten pages or less.

- Most sites used static web content – that is to say, all the information is present in the HTML data rather than being dynamically fetched and rendered at the moment the visitor arrives at their site.

- Sites often have a page with contact information on it, though it is common for this information to be present in some form on many pages, and occasionally there is no single page which includes all desired information.

- Many businesses have a related page on social platforms (Facebook, Instagram, Twitter), and a minority of them only have a social presence.

These valuable insights, which took a day to derive, allowed us to make quick, broad decisions regarding our initial implementation. In hindsight, we recognized some important oversights, such as a distinction between (large) chain businesses and the smaller "singleton" businesses. From a search perspective, chain data is of higher value. While chains represent a minority of actual businesses, they are perhaps the most important data segment because users tend to search for chain businesses the most. Chains tend to have more sophisticated sites, often requiring more sophisticated extraction technology. Extracting an address from plain HTML is far easier than extracting a set of entities dynamically placed on a map as the result of a zip code search.

Every Task Starts with Data   Developers can gain insights into the fundamentals of a domain by looking at a small sample of data (less than 100). If there is some aspect of the data that dominates the space, it is generally easy to identify. Best practices for reviewing data include randomizing your data (this helps to remove bias from your observations) and viewing the data in as native a form as possible (ideally seeing data in a form equivalent to how the machinery will view it; viewing should not transform it). To the extent possible, ensure that you are looking at data from production[5] (this is the only way to ensure that you have the chance to see issues in your end-to-end pipeline).

With our early, broad understanding of the data, we rapidly began building out an initial system. We took a pragmatic approach to architecture and infrastructure. We used existing infrastructure that was built for a large number of different information processing pipelines, and we adopted a simple sequential pipeline architecture that allowed us to build a small number of stages connected by a simple data schema. Specifically, we used Bing's cloud computation platform which is designed to run scripted processes that follow the MapReduce pattern on large quantities of data. We made no assumptions that the problem could best be delivered with this generic architecture, or that the infrastructure and the paradigms of computation that it supported were perfectly adapted to the problem space. The only requirement was that it was available and capable of running some processes at reasonable scale and would allow developers to iterate rapidly for the initial phase of the project.

Bias to Action   In general, taking some action (reviewing data, building a prototype, running an experiment) will always produce some information that is useful for the team to make progress. This contrasts with debating options, being intuitive about data, assuming that something is "obvious" about a data set, and so on. This principle, however, cannot be applied recklessly – the action itself must be well defined with a clear termination point and ideally a statement of how the product will be used to move things forward.

---

[5]"Production" refers to your production environment – where your product is running and generating and processing data.

If we think about the components needed for web mining local business sites to extract high-quality records representing the name, address, and phone number of these entities, we would need the following:

- A means of discovering the web sites

- A crawler to crawl these sites

- Extractors to locate the names, addresses, and phone numbers on these sites

- Logic to assemble these extracted elements into a record, or records for the site

- A pipeline implemented on some production infrastructure that can execute these components and integrate the results

Each of these five elements would require design iterations, testing, and so on. In taking an agile discovery approach to building the initial system, we instead addressed the preceding five elements with these five solutions:

- Used an existing corpus of business records with web sites to come up with an initial list of sites to extract from – no need to build a discovery system yet

- Removed the need to crawl data by processing data only found in the production web corpus of the web search engine already running on the infrastructure we were to adopt for the first phase of work

- Used an existing address extractor and built a simple phone number extractor and name extractor

- Implemented a naïve approach to assemble extracted elements into a record which we called "entification"

- Deployed the annotators and entification components using an existing script-based execution engine available to the larger Bing engineering team that had access to and was designed to scale over data in the web corpus

We discovered as part of this work that there were no existing off-the-shelf approaches to extracting the business name from web sites. This, then, became an initial focus for innovation. Later in the project, there were opportunities to improve other parts of the system, but the most important initial investment to make was name extraction.

By quickly focusing on critical areas of innovation and leveraging existing systems and naïve approaches elsewhere, we delivered the first version of the data in a very short time frame. This provided the team with a data set that offered a deeper understanding of the viability of the project. This data set could be read to understand the gap between our naïve architecture and commodity extractors and those that were required to deliver to the requirements of the overall project. Early on, we were able to think more deeply about the relationship between the type of data found in the wild and the specializations required in architecture and infrastructure to ensure a quality production system.

Going End to End with Off-the-Shelf Components   Look for existing pieces that can approximate a system so that you can assess the viability of the project, gain insight into the architecture required, recognize components that will need prioritized development, and discover where additional evaluation might be required for the inner loop. In large organizations, likely many of the components you need to assemble are already available. There are also many open source resources that can be used for both infrastructure and inference components.

Getting up and running quickly is something of a recursive strategy. The team makes high-level decisions about infrastructure choices (you want to avoid any long-term commitments to expensive infrastructure prior to validating the project) and architecture (the architecture will be iterated on and informed by the data and inference ecosystem). This allows the team to discover where innovation is required, at which point the process recurses.

To build the name extractor, we enlisted a distance learning approach. We had available to us a corpus of data with local entity records associated with web site URLs. To train a model to extract names from web sites, we used this data to automatically label web pages. Our approach was to

1. Randomly create a training set from the entire population of business-site pairs.

2. Crawl the first tier of pages for each URL.

3. Generate n-grams from the pages using a reasonable heuristic
    (e.g., any strings of 1–5 tokens found within an HTML element).

4. Label each n-gram as positive if they match the existing name
    of the business from the record and negative otherwise – some
    flexibility was required in this match.

5. Train a classifier to accept the positive cases and reject the
    negative cases.

If we think about the general approach to creating a classifier, this method allows
for the rapid creation of training data while avoiding the cost of manual labeling, for
example, creating and deploying tools, sourcing judges, creating labeling guidelines, and
so on.

Important Caveat to Commodity System Assembly   You may state at some
point that the code you are writing or the system you are designing or the platform
you are adopting is just a temporary measure and that once you have the "lay of
the land," you will redesign, or throw experimental code away, and build the real
system. To throw experimental code away requires a highly disciplined team and
good upward management skills. Once running, there is pressure to deliver to the
customer in a way that starts building dependencies not only on the output but
on the team's capacity to deliver more. Often this is at the cost of going back and
addressing the "temporary" components and designs that you used to get end to
end quickly.

Data Analysis for Planning

Now that we have a system in place, it's time to look at the initial output and get a sense
of where we are. There is a lot of value in having formal, managed systems that pipe data
through judges and deliver training data, or evaluation data, but that should never be
done prior to (or instead of) the development team getting intimate with every aspect of
the data (both input and output). To do so would miss some of the best opportunities for
the team to become familiar with the nuances of the domain.

There are two ways in which a data product can be viewed. The first looks at the *precision* of the data and answers the question "How good are the records coming out of the system?" This is a matter of sampling the output, manually comparing it to the input, and determining what the output should have been – did the machine get it right? The second approach focuses on the *recall* of the system – for all the input where we should have extracted something, how often did we do so and get it right?

The insight that we gained from reviewing the initial output validated the project and provided a backlog of work that led to our decision to expand the team significantly:

- Overall precision was promising but not at the level required.

- Name precision was lower than expected, and we needed to determine if the approach was right (and more features were needed) or if the approach was fundamentally flawed.

- The commodity address extractor was overly biased to precision, and so we missed far too many businesses because we failed to find the address.

- Our understanding of the domain was naïve, and through the exposure to the data, we now had a far deeper appreciation of the complexity of the world we needed to model. In particular, we started to build a new schema for the domain that included the notion of singleton businesses, business groups, and chains, as well as simple and complex businesses and sites. These concepts had a strong impact on our architecture and crawling requirements.

- The existing web index we leveraged to get going quickly was not sufficient for our scenario – it lacked coverage, but also failed to accurately capture the view of certain types of pages as a visitor to the site would experience them.

Now that we had a roughed-in system, we could use it to run some additional exploratory investigations to get an understanding of the broader data landscape that the project had opened up. We ran the system on the Web at large (rather than the subset of the Web associated with our existing corpus of business data records). More about this later.

Data Evaluation Best Practices   It is common for data engineering teams early on in their career to use ad hoc tools to view and judge data. For example, you might review the output of a component in plain text or in a spreadsheet program like Excel. It soon becomes apparent, however, that even for small data sets, efficiency can be gained when specialized tools that remove as much friction as possible for the workflows involved in data analysis are developed.

Consider reviewing the output of a process extracting data from the Web. If you viewed this output in Excel, you would have to copy the URL for the page and paste it into a browser, and then you would be able to compare the extraction with the original page. The act of copying and pasting can easily be the highest cost in the process. When the activities required to get data in front of you are more expensive than viewing the data itself, the team should consider building (or acquiring) a tool to remove this inefficiency. Teams are distinguished by their acknowledgement of the central importance of high-quality data productivity tools.

Another consideration is the activity of the evaluation process itself. Generally, there are judgment-based processes (a judge – that is to say, a human – will look at the output and make a decision on its correctness based on some documented guidelines) and data-based processes (a data set – often called a **ground truth set** – is constructed and can be used to fully automate the evaluation of a process output). Ground truth-based processes can be incredibly performant, allowing the dev team to essentially continuously evaluate at no cost.

Both tool development and ground truth data development involve real costs. It is a false economy to avoid these investments and attempt to get by with poorly developed tools and iteration loops that require manual evaluation.

Establishing Value

With the basics of the system in place, we next determined how to ship the data and measure its value, or impact. The data being extracted from the Web was intended for use in our larger local search system. This system in part created a corpus of local entities by ingesting around 100 feeds of data and merging the results through conflation/

merge (or record linkage as it is often termed). The conflation/merge system had its own machine learning models for determining which data to select at conflation/merge time, and so we wanted to do everything we could to ensure our web-extracted data got selected by this downstream system. We got an idea of the impact of our system by both measuring the quality of the data and how often our data got selected by the downstream system. The more attributes from our web-mined data stream were selected and surfaced to the user, the more impact it had. By establishing this downstream notion of impact, we could chart a path to continuous delivery of increasing impact – the idea being to have the web-mined data account for more and more of the selected content shown to users of the local search feature in Bing.

We considered several factors when establishing a quality bar for the data we were producing. First was the quality of existing data. Surprisingly, taking a random sample of business records from many of the broad-coverage feeds (i.e., those that cover many types of businesses vs. a vertical feed that covers specific types of businesses like restaurants) showed that the data from the broad-coverage feeds was generally substandard. We could have made this low bar our requirement for shipping. However, vertical and other specialized feeds tended to be of higher quality – around 93% precision per attribute. It made sense, then, to use these "boutique" vertical data sources as our benchmark for quality.

In addition, we considered the quality of the data in terms of how well that quality can be estimated by measurement. Many factors determine the accuracy of a measurement, with the most important being sample design and size and human error. These relate to the expense of the judgment (the bigger the sample, the more it costs to judge, the lower the desired error rate, the more judges one generally requires per Human Intelligence Task (HIT)[6]). Taking all this into account, we determined that a per attribute precision of 98% was the limit of practically measurable quality. In other words, we aimed to ship our data when it was of similar quality to the boutique feeds and set a general target of 98% precision for the name, address, and phone number fields in our feed.

---

[6]A Human Intelligence Task (HIT) is a unit of work requiring a human to make a judgment about a piece of data.

Because we had established a simple measurement of impact – the percentage of attributes selected from this feed by the downstream system – our baseline impact was 0%. We therefore set about analyzing the data to determine if there was a simple way to meet the precision goal with no immediate pressure on the volume of data that we would produce. The general strategy being to start small, with intentionally low coverage, but a high-quality feed. From there, we could work on delivering incremental value in two ways. The first involved extending the coverage through an iteration of gap analysis (identifying where we failed to deliver data and make the required improvements to ensure that we would succeed in those specific extraction scenarios). The second involved identifying additional properties of local businesses that we could deliver. Local business sites, depending on the type of business, have many interesting properties that benefit local search users such as business hours, amenities, menus, social media identities, services offered, and so on.

Discovering the Value Equilibrium   By establishing the evaluation process, the customer helped frame the notion of value. In many cases, declaring a target – say the data has to be 80% precise with 100% coverage – is acceptable, but a worthy customer would require more. In many projects, the data being delivered will be consumed by another part of a larger system and further inference will be done on it. This downstream process may alter the perceived quality of the data you are delivering and have implications for the goals being set. There is a tension between the lazy customer – for whom the simplest thing to do is to ask for perfect data – and the lazy developer, for whom the simplest thing to do is deliver baseline data. Setting a target is easy, but setting the right target requires an investment. The most important goal to determine is the form and quality of the output that will result in a positive impact downstream. This is the exact point at which the system delivers value. While managing this tension is a key part of the discussion between the customer and the team, it is even better to have the downstream customer engage in consuming the data before any initial goal is set, to better understand and continuously improve the true requirements.

Of course, determining the exact point at which value is achieved is an ideal. If the downstream consumer is unclear on *how* to achieve their end goal, then determining the point-of-value for your product would require them to complete their system. You have something of a data science "chicken and the egg" problem. In some cases, it is pragmatic to agree on some approximation of value and develop to that baseline. After this, the consumer can determine if it is enough for their needs.

There is one situation in which value in a data product can be relatively easily determined. If there is an existing product, then the customer can determine value in economic terms relative to the comparative quality of the data you provide. They may be happy to get slightly lower-quality data in return for lower costs, or pay more to get a significantly better product.

From Early to Continuous Delivery

We left our project, extracting local business data from the Web, open to delivering additional value in several ways:

1. Delivering more entities and covering the greater universe of local businesses in the market

2. Extracting more detailed attributes per business (such as images or reviews) and providing richer facets of information to users of the data

3. Applying the approach to different markets (including different languages), thereby extending the impact globally

4. Applying this approach to different verticals and extracting things like events or product information from a larger universe of web sites

5. Finding additional consumers of the data

6. Considering delivering the platform and tool chain itself as a product either internally or externally

Each avenue offers an opportunity to continuously deliver value to both the existing customer and parent corporation. The more we deliver to our immediate customer, the more we can positively and efficiently impact the product and our end users; the greater the ROI.

More Entities

Once we had established the feed as a viable data source for our first-party customer, we set about attacking the coverage problem from a couple of analytical angles. On one hand, specific implementation details and presentation formats of the web sites could make it easier or harder to extract the desired information. Take the address for example. If it was present in a single HTML node (such as a div or span) on the site and the site was presented in as a static HTML file, then extraction would be relatively simple. If, on the other hand, the site dynamically loaded content or used an image to present the address, then a more sophisticated technology would be required. These challenges necessitated successive investments in what we termed *technical escalation*. In other words, to get more data, we needed to increase the sophistication of our extraction stack.

Another consideration that our data analysis presented was the complexity of the underlying business. A small business with a single location would generally not present too much confusing information on their site. It would have a unique address, a single phone number, and so on. Now consider something marginally more interesting – a local bakery that participates in farmers' markets during the weekend. Now the site might legitimately include additional address information – the addresses of the markets where it would appear in the next few weeks. Fully exploring business complexity, we discovered that multi-brand, international chains not only present with multiple brands at many – potentially thousands – locations but through necessity leverage dynamic and interactive web site architecture to implement search engines for their customers to locate a chain location.

How to Look at Data   It is always important to view data in a form identical to that of your system's view, especially on the Web. If your system is processing web pages, reviewing example pages in a browser will hide many details, for example, there may be invisible portions of text in the underlying HTML. Viewing such a page in a browser will hide these, but your inference system will read these as being equivalent to the visible data. Only when your team has a full understanding of the underlying "natural" form of the data should you employ views that transform the data in some way.

With these two rich (and not entirely independent) dimensions to explore, we approached the planning by first prioritizing the potential investments in terms of expected impact and estimated cost. Second, we iterated on solutions to those areas in a tight loop: implement, evaluate, and improve.

In this way, we continuously improved our product by delivering more impact through coverage while iterating both on our overall architecture and on individual components.

The relationship between the concepts being modeled and the system requirements can be illustrated by our need to adapt the system to handle chain web sites. A chain, technically, is any business that has multiple locations for which the customer experience is interchangeable. An example would be McDonald's, because you get essentially the same experience at any location. Chains often make their location data available to site visitors through a search interaction, for example, you enter a zip code in the Starbucks store locator and get a response with a list of coffee houses in or near that location. To crawl and extract from these pages, we needed to emulate the user searching, dynamically render the result page, and follow the links from that page to individual store detail pages.

More Attributes

Another way we iterated on the value the system delivered was through additional attributes. Early on, we chose a design for the core of our system that included a simple interface for functions we called *annotators*. An annotator is the computational analogy of a human annotator who looks at a page and highlights spans of text that denote a specific type of thing. We created name annotators, address annotators, phone annotators, and so on. Each could be implemented in complete isolation, tested, evaluated, and deployed to the pipeline through a configuration – we call this approach a plugin architecture (see Figure 1-1 for an overview).

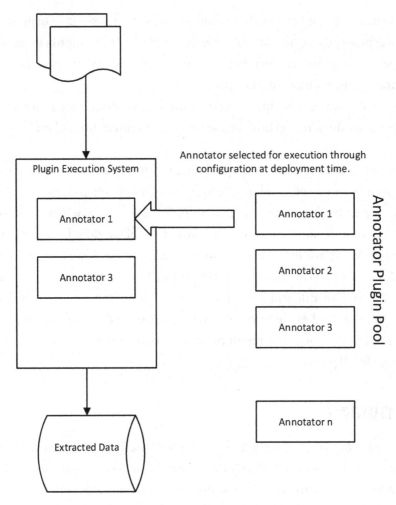

Figure 1-1. *In a plugin architecture, components – in this case, annotators – can be developed, tested, and managed independently of the runtime. A configuration is used to deploy components to a pipeline as required.*

Isolation allowed us to work on an annotator, without dependency on other components, and deploy with no additional testing to other components. In this way, we could expand from our original three properties – name, address, phone – to an arbitrary set of properties. Perhaps the most interesting of these was the business hours annotator. Understanding how this was developed provides some interesting insights into the relationship between the nature of the entities in the real world, the manner in which they present themselves on the Web, and the requirements of a system designed to extract that data in the context of a local search engine.

It is possible to model business hours semantically by intuition – many businesses have a simple representation of open and closed times, one for each day of the week. But it requires looking at the data, that is, actual business web sites, to truly understand the sophistication required to both model the domain and build the right extraction technology. Businesses may use inclusive and exclusive expressions ("open from 9 to 5 but closed from 12 to 1 for lunch"), they may be vague ("open until sundown"), and they may describe exceptions ("except for national holidays"). In more sophisticated sites, business hours may present in the moment, that is, in a way that only makes sense in the context of the date the site is visited ("Today: 9-5, Tomorrow: 9-6, etc.).

Studying the data led us to think hard about how we collect data in the first place through crawling the Web. While the general web index was fine as a starting point, if just 1% of businesses in our corpus were changing their business hours every week, we would need to refresh our data far more frequently than that of a large-scale index which would not update most business pages as quickly as it would update a more dynamic page like that of a news site or social site. In reality, the number of sites required to service a local business extraction system is many orders of magnitude smaller than that required for the general web search engine. And so, it made perfect sense for us to remove our dependency on this initial approach and invest in our own crawling schedule.

Finally, business hours are not linguistically similar to names, addresses, or phone numbers, having far more internal structure and variance of expression. This meant that we had to invest more heavily in the specific annotator to deliver business hours data. Again, studying the data informs the next iteration of architecture, requirements, and technology choices.

More Markets

As the project progressed, we began connecting with other teams that were interested in the general proposition – leveraging the Web as an alternative to licensed feeds for local search. This led to partnerships with teams tasked with delivering local data for other markets (e.g., a country-language pair). By taking the pluggable approach to architecture and partnering with other teams, we transitioned from being a *solution* for a single scenario to a *platform* for a growing number of scenarios.

Different markets presented different motivations and variables. In some cases, less budget for licensed data incentivized the team for that market to seek out more automated approaches to delivering data; in others, there simply weren't any viable sources of data approaching the quality that the end product demanded. To continue delivering on the investment at this level required us to fully support the operational side of the system (from deployment automation to DRI[7] duties). We provided approaches to deliver the pluggable components in a way that could leverage the experience in the original instantiation, but was also flexible enough to meet the specific needs of those markets, and a commitment to creating, maintaining, and improving tools for every part of the developer workflow.

More Quality

Expanding the coverage of a system requires constantly assessing and maintaining the quality of the output. In some cases, this can involve adjusting and evolving the current method. In others, it requires investing in a switch to a more general approach or some sort of hybrid or ensemble approach. In our journey with local extraction from the Web, we encountered many instances of this.

We wanted to switch the method we were using for address extraction. The initial commodity solution, which was giving us reasonable precision but lacked recall, had to be replaced. We opted for a machine-learned approach (based on sequence labeling) and an efficient learning method similar to active learning. To make the transition, we created a regression set and continuously improved the model until we were satisfied that it was a viable replacement. From there, we could take advantage of a more general model to continue to extend our recall and maintain quality.

Another very real aspect of shipping industrial inference systems is managing errors that impact the current performance or are otherwise of great importance to one of your consumers. A typical problem with extracting from the Web is that the distribution of problems in the population changes over time. For example, while the system might have happily been delivering data from a specific site for some time, if the site makes a change to a presentation format that wasn't present or was rare in the development period, the system may suddenly stop producing quality results, or any results at all. If the consuming system was, for some reason, very sensitive to the data from this site, then it would be pragmatic to implement either some form of protection or some form of rapid mitigation process.

---

[7]A DRI is an acronym for "Designated Responsible Individual." This is the person who is on call on a particular week to resolve any live site issues that may come up with the running product.

The Platform as a Product: More Verticals and Customers

Having expanded the contributions made by the investment in web mining through coverage, attributes, and markets, the future for additional value lies in transforming the platform itself into a product. Getting to this level requires a decision around the business model for the product. While our original goal was to deliver business data, expanding to ship the platform as a general-purpose capability meant that we would be shipping software and services. This path – from solution to platform to product – is not uncommon and represents orders-of-magnitude jumps in value of investment at each step.

Early and Continuous Delivery of Value

Projects that aim to extract value from data through some form of inference are often, essentially, research projects. Research projects involve many unknowns and come with certain types of risk. These unknowns and risks tend to be markedly different from those in traditional application development. Consequently, we need to rethink our notions of the customer, the core development iteration, what form our product takes, and how we deliver.

At the core of the types of projects we are interested in is some type of inference over potentially large data. Your team must have the ability to intelligently explore the search space of data and inference solutions in the context of an overall product and well-defined business need, so that

- The customer is engaged, and their expectations are managed – an engaged customer will ensure that the details of the final product align with the business case requirements. The customer will see the implication of the project – especially how the nature and potential of the data can influence the final result and be best consumed.

- Progress is articulated appropriately and continuously – including prototypes, preliminary improvements that are below the required standard for a Minimal Viable Product (MVP), and eventual success in delivering and maintaining the data product. An ideal customer will even get ahead of delivery and assist in the success of the project by building stubs or prototypes of the consuming system if they are not already available.

- The customer fulfills the role of helping the team to establish and prioritize work. The customer understands that involvement is a vital and equal voice that helps guide, shape, and land the project.

Projects, both traditional and data, tend to have the following phases:

1. Initialization

2. Delivery of a baseline product, or Minimal Viable Product (MVP)

3. Delivery to requirements

4. Maintenance

5. Expansion

In most of the phases, the central development loop, that is, the process by which your team will deliver some sort of inference capability, will be characterized by the workflow illustrated in Figure 1-2.

- Requirements generation is an iterative process involving the customer and the development team, refining the initial tasks through analysis of data representative of the population targeted by the system.

- An approach to evaluation is determined and implemented to allow progress toward the targeted data quality to be tracked.

- The development team collects data to be used in developing inference components, including training data and evaluation data for their inner loop.

- Evaluation, data analysis, and error analysis are used to improve the models but also, where needed, to improve the evaluation process itself.

Our goal in this book is to help you navigate delivery through this workflow and to understand how the contexts of each process influence the overall success.

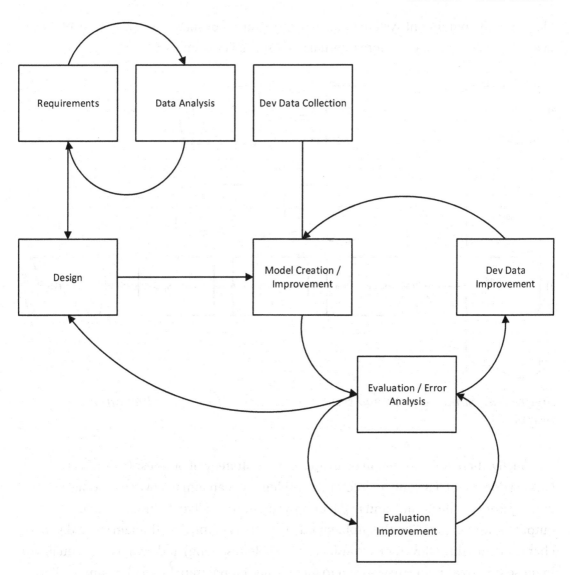

Figure 1-2. *The process by which a team delivers inference capability*

While a lot of focus is given to advances in the core models and training techniques in the inner loop, in industrial contexts, it is also important to acknowledge the position your contribution will have in the overall workflow of the deployed system. As shown in Figure 1-3, you may be responsible for a simple, single-component system or be building a larger, multicomponent system or developing a subset of such a system. In any of these cases, you may have one or more internal and external customers.[8]

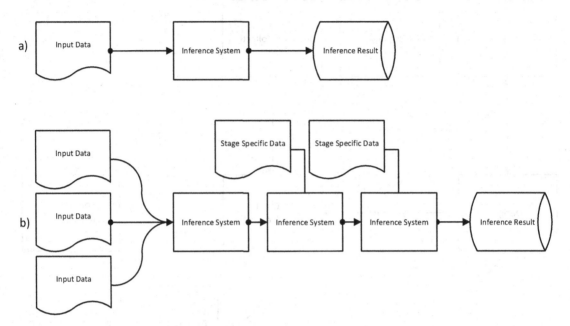

Figure 1-3. *Simple end-to-end system (a) vs. a multi-data, multicomponent system (b)*

Finally, there is the selection of an appropriate strategy of progress (Figure 1-4). Given a domain of data and an inference problem, we can aim to cover the domain entirely from the beginning and make improvements overall to the quality of our output, or we can partition the problem into sub-areas of the data domain (with distinct characteristics that allow for narrow focus in problem solving) and achieve high quality in those sub-areas before moving on to include additional tranches of the domain. For example, if we were building a system to identify different types of wildlife in video, we could start with building a data set for all types of animals and then training our system to recognize them. For a fixed cost, we would have a small amount of data for

---

[8]The notion of internal and external is somewhat arbitrary, especially in large corporations. The reality is that some customers are closer to you organizationally and some are at a greater distance.

each animal, and our recognition quality might be low. For the same cost, we could alternatively create labeled data for just a small number of animals and get quickly end to end with high quality.

Another example of this is in the domain of information extraction from the Web. A "quality first" strategy would focus on simple web sites where there is little variation between sites and few technical hurdles (e.g., we could limit our input to plain HTML and ignore, for the time being, problems associated with dynamically rendered pages – images, flash, embedded content, etc.). A "breadth first" strategy would attempt to extract from all these forms of input with the initial system and likely end up with lower-quality extractions in the output.

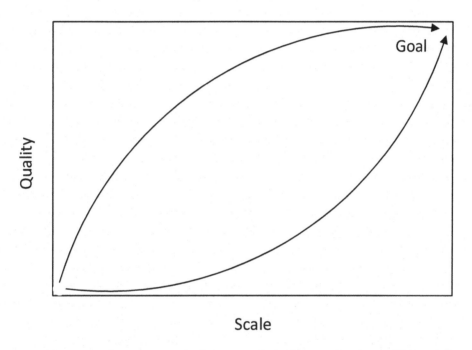

Scale

Figure 1-4. *Strategy for progress in large-scale inference projects. The top line represents prioritizing quality first and then turning attention to scale. Here, scale means the ability to handle more and more types of input. In a document processing system, for example, this would mean processing many genres of document in many file formats. The bottom line represents prioritizing scale first and then working on quality – in other words, investing in handling a large variety of documents and later driving for quality. In an agile process, the top line is preferred as it is the shortest route to delivering value to the customer.*

Conclusion

In Chapter 1, "Early Delivery", we looked at ways in which a team can get up and running on a new inference project. We touched on the central role of metrics and a culture that is biased to action to enable quick iterations using data analysis to inform the evolution of all aspects of the project.

In Chapter 2, "Changing Requirements", we will discuss how you can build systems that are designed for change.

Changing Requirements

*Welcome **changing requirements**, even **late in development**. Agile processes harness change for the customer's **competitive advantage**.*

—agilemanifesto.org/principles

The one constant in data engineering projects is change. Sometimes the change is in the data itself – for example, a system to detect fraudulent patterns of data will inevitably be outsmarted by future attacks causing the system to become less effective. Ideas about what to do to solve a data problem change as the data is explored and analyzed. Customers can be fickle and change their mind about what they want and what features are most important to them. There is also change in the machine learning algorithms and techniques available to use – machine learning research is continually moving forward. Finally, competing offerings are changing – a strong competitor is often difficult to keep up with and must be tracked closely to ensure they aren't getting an insurmountable competitive advantage.

The most successful data engineering projects consider change to be inevitable and build in at the start mechanisms for dealing with change. In this chapter, we will consider several strategies for dealing with change which include building models that are more resilient to change, monitoring change in the performance of shipped features and models, ensuring that models and their features are testable and agile, and measuring the performance of your own product and competing offerings.

Building for Change

Perhaps the hardest culture to develop in a data engineering team, especially one that is just starting out, is the idea that "we are in this for the long term." If you don't start with that cultural principle, the investments that are made will all be short-term wins that will be increasingly difficult to maintain over time as requirements and the data landscape change.

© Eric Carter, Matthew Hurst 2019
E. Carter and M. Hurst, *Agile Machine Learning*, https://doi.org/10.1007/978-1-4842-5107-2_2

It is also important to not get too attached to a particular component or architecture. If systems are designed in a loosely coupled way from the start, rewriting a problematic component or system should always be considered as an alternative to refactoring or continuing to fix a long tail of bugs in that component. As discussed in Chapter 1: Early Delivery, this provides a strategy for getting started – connect a lot of existing components together to get something working end to end quickly, but do it in a loosely coupled way so you can replace or upgrade parts of the system as needed. You should be ready to throw away the first system you build as you learn more about the problem you are trying to solve and find that the initial pipeline of components doesn't work well for the problem at hand.

Measurement Built for Change

As mentioned in Chapter 1: Early Delivery, one of the most important things to establish early on in a team is metrics. Once metrics are established, you must fund a dedicated effort to keep the metrics aligned to the business goals of the project. We found that on a project with about 50 developers, it was a good idea to budget five developers to focus solely on measurement. At times it felt like even that was too little of an investment. We called this team of five developers the "Measurement Team."

Measurement   Measurement is to data projects what unit tests are to traditional software projects. If you have robust measurement, you can evaluate new components, new architectures, and new paradigms against a truth. If you don't take measurement seriously and invest in it extensively, you can never have a successful data engineering project.

What did these five developers do? Well, first they provided a firewall for the rest of the team to prevent the rest of the development team (and shipping models) from being contaminated by test sets. One key principle to follow in machine learning is to not contaminate your training data with your test data or developers will start to learn patterns in the test data, overfit to the problem being measured, and skew the metrics so it looks like you are doing better than you really are. Most significant data problems will be apparent in your training data.

We built an organizational boundary around this principle by keeping the test sets only accessible to the measurement team. In practice, we found it was too easy to have test data leak into training data or subtly into developers' minds and effectively fool ourselves into thinking we were better than we were without this organizational boundary in place. The measurement team also acted as the authority for the official key metrics for the team which were calculated and tracked on a regular basis and reported out in a portal.

The measurement team created systems that could calculate and track any number of data metrics the team would need and helped to maintain and store the hundreds of different test sets that the team would need. From the start they built measurement systems anticipating that the training and test sets would change and the team metrics would change.

They also designed the web-based applications that human judges would use to score and label data and systems to manage and track judgments made on the data. Some of these judgments were used to evaluate our performance on certain key metrics – for example, the accuracy of business phone numbers. Other judgments were used as training data for machine learning models.

The labeling web applications were called HIT apps. A HIT (Human Intelligence Task) was a judgment made by a judge about a piece of data, and the system was designed to maximize the number of judgments a judge could make per hour while simultaneously maximizing the accuracy of each judgment. HIT apps were built on a general platform that served many similar teams and ensured that each team didn't have to reinvent common functionality used in our scoring and labeling systems. These systems were designed with the expectation that the type of HIT apps needed would change over time and even that strategies for managing judges would change over time.

Invest in Tools   Many teams underinvest in tools that in the long run will give the team great efficiency. Just as a team writing code benefits greatly from a high-quality development IDE, source code control, continuous integration and deployment system, and so on, a data engineering team benefits greatly from high-quality machine learning tools, HIT app creation tools, and data exploration and evaluation tools.

A team creates systems for measurement very differently if they are building them for the "long term." For example, it becomes important to have a HIT app system with reusable components that can be quickly leveraged when the team decides they need to not only measure the accuracy of business phone numbers but also phone numbers of associated departments within more complex businesses. A system to track training sets and test sets is built much differently if from day one you assume you will potentially have thousands of these sets in the long term. Systems to calculate metrics are designed much differently if the team knows they will eventually have to report hundreds of different metrics for multiple different markets – for example, not just phone number accuracy in the United States, but also for 30 additional countries.

The investment in robust systems like this certainly can slow down initial work – but in the long run, the increased velocity more than pays for the increased cost of designing for change from the start.

Is "Building for the Long Term" Just Waterfall?   Building for the long term sounds suspiciously like waterfall development – the approach of figuring out everything you need to build up front before developing seems anathema to Agile. What we are talking about here is maybe better expressed as **pick flexible architecture patterns that enable long-term change**. As an example, we didn't immediately build support for 30 markets – we initially built support for just the US market. But wherever we could in the code, we ensured we didn't hard code to one market as we knew we had to support many. When then in later iterations we began to support our second and third and fourth markets, changes had to be made to the existing code, but those changes weren't as extensive as they would have been had we not at least planned with our architecture patterns to support multiple markets.

Pipelines Built for Change

A second large investment we made in our team of 50 was to fund a pipeline and infrastructure team. This team provided the infrastructure to deploy and run all the models and code that the rest of the team would use to generate a new data catalog every day. In a team of 50, we dedicated ten members of the team to pipelines. We called this team the "Data Pipeline" team.

The data pipeline team created the online system or "pipeline" that would ingest all updates to our disparate data sources every day, run code to normalize that data, host a variety of models to improve and aggregate that data, host a stage of our pipeline that we called "conflation/merge" which was how we deduplicated the many data sources that we had that all had information about the same entity in the world and tried to pick the best data from the multiple sources available, and then ultimately publish the data to our catalog.

Environments or Rings   It is beneficial for teams to deploy their system to multiple environments (also sometimes called rings) that progressively increase in stability. At Microsoft, we generally had three environments. The "Dev" environment had the version of the pipeline with the latest check-ins where developers worked actively on the product. Loads on the pipeline in the "Dev" environment represented only a fraction of the real load the system needed to work with. As code in the "Dev" environment was stabilized and all tests passed and pipelines were verified to work, the code would then be moved to the "PPE" or pre-production environment. This environment would take on loads more representative of production loads and acted as a final smoke test environment before moving code to "Prod" or the production environment. We established "gates" between the environments – basically a battery of tests and metrics that needed to pass and be verified to not regress before code could move from Dev to PPE and then from PPE to Prod.

Figure 2-1. *A data pipeline*

As a brief explainer for our system as we will discuss aspects of it throughout this book, our pipeline as shown in Figure 2-1 ingested data about local businesses from hundreds of "feeds" that were provided by multiple data providers across the world. For example, we had feeds that came from the system we built to extract local businesses from primary web sites on the Web which is discussed in Chapter 1: Early Delivery – for example, we would crawl the Walmart site to get the location, phone number, and other information for every Walmart in the world. We also had feeds from third-party data providers like Yelp and TripAdvisor that had additional data about businesses such as reviews. We also had feeds from third parties that scanned phone books. Each day, all that data was re-fetched and ingested into our data pipeline. The data was normalized (e.g., capitalization was standardized, phone number formats were standardized, etc.). We then geocoded all address information to a latitude/longitude point on a map using our geocoder.

The data was then matched using the conflation/merge record linkage system, which will be described later in this chapter, consisting of four steps: candidate set generation, matching, clustering, and merging. The conflation system could detect that record 9875 in the TripAdvisor feed was talking about the same business as record 5324 in the Yelp feed and record 1945 in the web-extracted feed. These three records (in reality, we would often have between 25 and 100 feeds that all had a corresponding record to contribute to the more popular businesses) would need to be merged together – each provider had subtle differences in their data, so we had to use machine learning to both decide if two records matched and were talking about the same business and decide whether to use the phone number provided by Yelp, TripAdvisor, or web-extracted data when the providers had different opinions about what the phone number was. The merged entity was then published to our catalog.

The team created the data pipeline very differently knowing it was going to need to process multiple markets worth of data, that it was going to be used by many different teams in Microsoft, that other teams would need to extend and configure the pipeline in different ways, that data feeds and data sources could be modified and added and removed at any time subject to current business deals and available data, and that a mechanism was needed to manually correct data and have a "correction" layer to override any bad results the data pipeline might produce for a particular entity.

Building for change, extensibility, and configurability did involve more upfront work in the beginning; but our long-term goal was to ensure that Bing's local business catalog was competitive with Google through leveraging a variety of data sources, the Web, and advanced ML models. We wanted to be competitive with Google in every country, not just the United States. So the data pipelines were designed in a way that they were very configurable and extensible from the start. Pipeline monitoring and tools were designed to anticipate multiple data pipelines – one for each market which could run independently from one another. Later as we brought online new countries, the process was very quick because we had designed with long-term goals in mind.

Support Teams for Machine Learning   So far, we have basically described two disciplines within our team – a measurement discipline and a big data pipeline discipline. Both disciplines had to be conversant with machine learning to ensure that measurement and the data pipeline supported the data scientists creating the models that would run in production. However, this often led to interesting management challenges. Often, people in the "measurement" discipline or the "big data pipeline discipline" looked across the org at the developers who were in the "data scientist" discipline and building the production models and would say, "I want to do that." In general, we would encourage and support people who wanted to do that as much as possible as it helped the entire team get better as knowledge from the different disciplines moved throughout the various teams. It helped the data scientists understand "what could be so hard about running a data pipeline" when someone with that background worked in their area. Curiously though, we rarely had data scientists ask to spend more time working on the pipelines. To offset the lack of desire of data scientists to work in pipelines, we often required data scientists to go further once they had a model working on their machine to not just throw the model over the fence to the data pipeline team and say "Please run this" but actually have the data scientists also do the work to check in and integrate their model to the data pipeline and ensure it ran well in production.

Models Built for Change

In this section, we will look at how machine learning models can be built for change. The example we will consider in this section is Bing's conflation/merge system which was used to link records provided from multiple data providers to create each local entity.

Introduction to a Conflation System

A third large investment we made in our team of 50 was to fund the conflation/merge team. This team owned the complex algorithms and machine learning models that were used to do record linkage for the system and pick the best values for each key attribute of a business when potentially dozens of different values were available to pick from. This was the most complex area of our system and had another ten developers dedicated to it. We called this team of ten the "Conflation" team.

The conflation team created the system that matched data coming from the many different local data providers that Bing had licensed data from in addition to data we extracted from the Web. This record linkage system followed the classic model described in Chapter 1: Early Delivery: it started simple, and then as requirements and measurements evolved, the system evolved from a simple rule-based system to a much more complex ML-based system.

The record linkage component was identified as a key area for improvement and rewrite once metrics were developed that showed weaknesses in our record linkage system. For the end user, these weaknesses in our system surfaced as "duplicates" – for example, a user might search for Starbucks and where the ground truth was that there were five Starbucks in their neighborhood, the search engine might come back with six Starbucks. This error would typically be caused by record linkage errors where a record that corresponded to the fifth Starbucks in the list was not properly linked to the fifth Starbucks due to the variation in the data in the records being too high, so instead it became an erroneous sixth entry.

Figure 2-2 shows an actual example of this error that comes from Bing in 2014. These two entities were constructed from a set of 18 records from 11 providers that should have all been linked together to create a single theater entity. But our record linkage system saw enough records out of the 18 that were different enough from the others that it decided to create two entities. This happened because of the differences in name, phone number, and reviews coming from different providers. All the 18 records and 11 providers associated with this entity are shown in Table 2-1.

Figure 2-2. *An "undermatch" or duplicate error in the data from Bing in 2014*

Table 2-1. Multiple data providers and records used to create one entity

Provider	Name	Closed?	Address	Phone	Web Site
A	Village Theatre		480 1st Ave NW, Issaquah	4253950051	Villagetheatre.org
A	Village Theatre		215 Front St N, Issaquah	4253918190	Villagetheatre.org
B	Village Theatre		303 Front St N, Issaquah	4253922202	Villagetheatre.org
C	Village Theatre		303 Front St N, Issaquah	4253922202	Villagetheatre.org
C	Francis J Gaudette Theatre		303 Front St N, Issaquah	4253922202	Villagetheatre.org
D	Francis J Gaudette Theatre		303 Front St N, Issaquah	4255392202	
D	The Village Theatre		215 Front St N, Issaquah	4253918190	Villagetheatre.org
D	Village Theatre		303 Front St N, Issaquah	4253921942	
D	Village Theatre		480 1st Ave NW, Issaquah	4253950051	
E	Village Theatre	False	303 Front St N, Issaquah	4253922202	Villagetheatre.org
F	Village Theatre		303 Front N St, Issaquah	4253922202	
F	Francis J Gaudette Theatre		303 Front N St, Issaquah	4253922202	
G	Village Theatre		303 Front St N, Issaquah	+1 425-392-2202	Villagetheatre.org
G	Francis J. Gaudette Theatre		303 Front St N, Issaquah	+1 425 392-1942	
H	Village Theatre		303 Front St N, Issaquah	4253922202	Villagetheatre.org
I	Village Theatre		303 Front St N, Issaquah	4253922202	Villagetheatre.org
J	Village Theatre		303 Front St N, Issaquah	4253922202	
K	Village Theatre		303 Front St N, Issaquah	4253922202	VillageTheatre.org

Bad Data Proliferates   You may wonder why multiple data providers out of 11 in the preceding example had identical bad data. This issue happened frequently in our space because of the buying and selling of data sets that occurs in the domain. A single data provider might get a bad phone number or name (or more likely an old and out-of-date phone number). That data provider might sell their data to other data providers. Those data providers typically augment the base data by adding reviews or some other value on top. They in turn resell their data passing on the bad phone number or name. Bing as a consumer of multiple of these purchased data sets would often have multiple providers that seemed to "confirm" an entity existed because of their agreement in data, but in reality the seeming "confirmation" that this must be a real entity because multiple providers were describing it in the same way was actually bad data shared among data providers.

Real-World Data Is Noisy   As you study Table 2-1 which shows the actual name, address, closed status, phone number, and web site from 11 providers, you will see a lot of variation for each field. There are three choices for names for the entity, four different address choices, four choices for phone number once it is normalized, and multiple gaps in what each provider knows. For example, the web site is only known 11 out of 18 times in this example, and only one provider is confident enough to assert the entity is open.

Also, you will note a lot of duplicate data provided by certain providers – for example, Provider D has four records all purporting to be the Village Theatre.

As users began to report and complain about these duplicate entities in our system, we began to develop a measurement that we called "duplicate rate" to measure the problem. We would have human judges run a corpus of queries representative of actual queries being made by users on the site and then judge the results of each query to see how many duplicates were produced. As you might imagine, the judging process was complex as it often isn't obvious that two returned values are duplicates. Judges would have to look more deeply at the records we linked together, make phone calls to businesses, or try to find primary web sites to validate that a particular entity was a duplicate or not.

Now that we had a duplicate rate metric developed and measured on a regular basis (we would update it about every 2 weeks or more frequently when the engineering team needed more frequent updates), we began the process of trying to improve it. The team also created extensive labeling of records to indicate which records corresponded to a real-world entity. This effectively represented a human doing manually what we eventually wanted the system to do. The way we did this was we took all the records for a given zip code – which often had dozens of records from different providers representing the same business – and manually clustered together these records to generate the training data that was used by the system.

The team then used this training data to use more sophisticated techniques for record linkage and move beyond a rule-based system. The team had to match, cluster, and then merge together multiple pieces of data about the same business provided by multiple feeds from different data providers. The team created a machine-learned match function that was used in matching in addition to machine-learned models to merge data together and sophisticated algorithms to cluster related data together into the same cluster. The corpus of human-labeled training data representing record clusters was used to determine new features that could be added to the match model or cluster and merge models to ensure that our duplicate rate decreased.

As our match model improved and our duplicate rate went down, users started to complain about a new problem. The way this problem usually manifested itself was a particular entity would have a bad phone number, address, or associated reviews. When looking into the issue, we found that our system now was too aggressive at reducing duplicate records and was now creating what we called "overmatches" where records were linked together for distinct businesses that were close together (such as two Starbucks within a block from each other which often happens in Seattle) or a business that was contained within another business like a coffee shop inside a supermarket.

Figure 2-3 shows an example of an overmatch. In this case, a record for a "Levi Store" got linked with a record for a mall called "Seattle Premium Outlets." Then the merge system would determine which data from all the records were linked together and believed to represent a single entity. Based on that determination, it would construct the final entity by taking the name from a record provided by one data provider – predicted to have the best names – and the web site from a record provided by a second data provider, predicted to have the highest-quality web site. Unfortunately, the record provided by the first data provider actually represented the real-world entity "Seattle Premium Outlets," and the record provided by the second data provider actually represented the real-world entity "The Levi's Outlet."

Figure 2-3. *An overmatch in the data – from Bing in 2014*

In response to overmatch problems, we developed a second "overmatch" metric that we began to regularly measure in addition to duplicate rate. Now we asked judges to run a set of queries and look at records the system picked to create each entity. Over time, we learned we had to fine-tune the system to optimize both our duplicate rate and overmatch metrics. Often, the work required to improve duplicate rate could regress overmatch. Sometimes specific work was required to reduce overmatch that didn't directly impact overmatch rate. For example, one common cause of overmatch is the preceding example of a parent entity "Seattle Premium Outlets" being conflated with child entities, the "Levi Store" in the outlet mall. Work to reduce this overmatch typically ended up creating new entities in the catalog that were previously submerged in overmatched entities rather than changing the duplicate rate.

The Conflation System

We will briefly consider the steps that the conflation system used to link records together and ultimately merge those records to create each local business entity in the system in Figure 2-4. The system consumed data provided by multiple data providers which had already been normalized.

Figure 2-4. *The conflation system*

1. Candidate set generation: In conflation, the fundamental operation is to look at two records in the system and decide how similar or dissimilar they are. This is accomplished by a machine-learned function we called the "match function" that takes as input all the attributes of each entity along with many calculated features based on the attributes of each entity and outputs a similarity score from 0 to 1. The match function is expensive to run – so there is no way you can run it on the entire n x n matrix of records in the system when n is in the hundreds of millions of records for our system. It also doesn't make sense to do so – it is wasted Compute time to compare records that are clearly talking about a McDonald's in Topeka, Kansas, with records that are talking about a McDonald's in Boise, Idaho.

So the first stage of our conflation system looks through the records in the system and generates smaller sets of records (which we called candidate sets) which have a high likelihood of being in the same physical location. For example, a naïve implementation of this would be to select the set of all the records that are in a particular zip code and run pairwise the match function on all those records.

In reality, we divided the map into smaller tiles we called quadkeys (roughly the size of a city block). Any entity in a given quadkey would be compared with all businesses in the same quadkey and the eight other quadkeys surrounding that quadkey as shown in Figure 2-5.

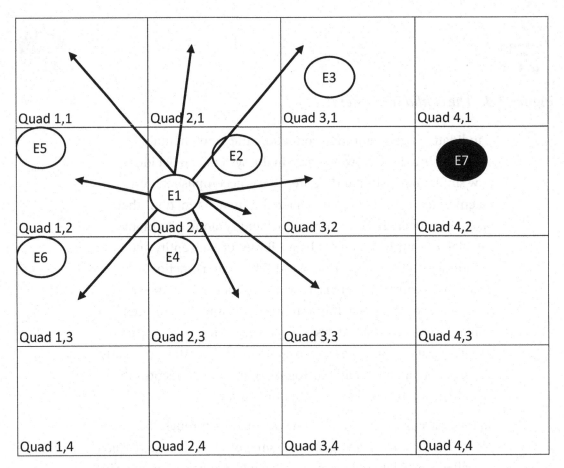

Figure 2-5. *Quadkeys – an entity (E1) is compared with all entities in the same quadkey and the eight adjoining quadkeys. All entities except E7 are compared by this algorithm.*

2. Match: At this stage, the match function which took as input hundreds of features that were inferred and calculated for an entity pair would be run doing an n x n comparison of all entities within a candidate set. As a further optimization, we had a lower-cost match function that could be run initially to rule out obvious non-matching entities and then a higher-cost match function that could be run on entities that the low-cost match function could not rule out as being similar or dissimilar.

For example, we invested in a set of natural language processing (NLP) features that would take the input name of an entity and for each word in the entity name would tag words that were determined to be a person name [P], the name of a business [B], the category of a business [C], the department of a business [D], and so on. For a business called "State Farm Insurance – John Smith," the tagger would tag it as "State[B] Farm[B] Insurance[C] – John[P] Smith[P]." From these additional inferred tags, we created features that the match function learned that took such as "Both Entities Have Person Names" or "Person Name Similarity Score" that would take as input the tagging data from both entities being compared and output a 0-1 score.

The match function was then trained against data labeled by judges where a judge would determine whether two records from data providers were talking about the same entity (see Table 2-1 for an example of data coming from data providers) and the match function would learn which of the hundreds of features it was looking at such as "Both Entities Have Person Names" or "Both Entities Have the Same Phone Number" should be weighted highest when comparing entities of various types.

3.   Connect and cluster: At this stage in the pipe, the match function now has output similarity scores for all pairs of entities that are nearby each other. This yielded what we called a "connected component" where all entities for which the match function was run now have an edge between them. With entities connected, we now have a new problem to solve, a problem we called "clustering" – which is how to determine which records should be combined to create a particular cluster which will in turn yield a local business.

Clustering determines how many entities to create from that connected graph. For example, Figure 2-6 shows how clustering would operate on a simple connected graph. Records are represented by the nodes a through f. Match scores are on the edges between records. The graph is not completely connected because the records a, b, and c were in a different candidate set than d, e, and f (likely because the records are too far away from each other). In this example, three entities are created from the clusters: abc, e, and df. The match scores (on edges between records) that are greater than .5 are considered a match and match scores lower than .5 are considered a no match.

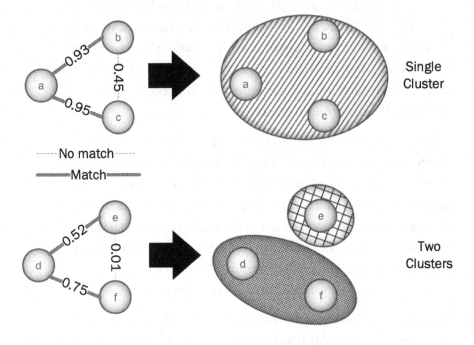

Figure 2-6. *Creating clusters from a connected graph*

4. Cluster correction: At this stage, we kept lists of known duplicates or overmatches in the system that were reported by users that the conflation system still wasn't getting correct. So we would have "forced matches" to bring together duplicates that the system failed on and "forced unmatches" to force apart entities that were overmatched. Over time, as our system improved, less of these manual corrections were needed.

5. ID enrichment: We will discuss this later in this chapter when we talk about issues around ID stability.

6. Merge: At merge time, we would take all the records that were linked together in a cluster and make a decision as to which data we would use from all the records in the cluster to create the final entity. Here too, we used machine learning to determine which data provider had the most correct or trustworthy data to pick. For example, in a cluster with data for a restaurant, the machine-learned merge function might learn to trust a restaurant-specific provider for the name for the entity but to trust a phone book–scanning provider for the phone number for the entity.

7. Merge correction: At this stage, we kept lists of known merge failures reported by users (e.g., cases when we picked an attribute that was bad data from a particular provider but another provider provided a correct value for that attribute). We would then force picking the data from the provider known to be correct. If none of our providers were correct, we also had our own Microsoft-provided feed of data that would cluster into an entity that we had manually curated and knew to be correct.

Building the Conflation System for Change

Perhaps the most significant way this team designed the conflation system for change was through anticipating that one match function would not work equally well for all world markets. For example, the match function for the US market is highly dependent on accurate addresses and the latitude and longitude for businesses. But in a market like Brazil, addresses are much less structured and vague than in the United States – for example, Brazil addresses will often only indicate a neighborhood with a location relative to some landmark in the neighborhood. By anticipating that multiple match functions would be needed, the team built the conflation system in a way that many elements could be replaced with different more market-specific elements – for example, the candidate generation system that in the United States was highly dependent on address could be swapped with a less address-sensitive candidate generation system for countries like Brazil.

Also, with the realization that many different match functions would need to be built, the team built a library of reusable features that could be leveraged in each market-specific model. This library of reusable features with support for calculating name similarity, address similarity, category similarity, phone number similarity, number of spaces in a string, punctuation, and named entity recognizers could be leveraged by the multiple conflation models that were created over time, thereby allowing the team to be robust in the face of changing requirements. Also, part of this library of reusable features were reusable tools for generating and processing the training data for each feature.

An Architecture to Enable Change

One tough decision the team made early on was to fundamentally pick an architecture for our data pipeline and data catalog that enabled change. That architecture decision was referred to as the "build the world from scratch every day" decision. There were pros and cons to this decision. We will describe what this decision was, why it was made, how it helped us, and how it hurt us.

Early on in the process of creating the Bing local data pipeline, it was clear to us that our conflation models were still raw and immature and we had a long ways to go to achieve the duplicate rate and overmatch rates we hoped for. Unlike Google, we didn't have an army of people at our beck and call – both paid and unpaid – who would correct all the errors in our catalog for us. We decided to bet very heavily on machine learning and extracting business entities from the Web rather than human curation and correction. For that reason, we decided that we would effectively throw away our catalog every night and build it from scratch the next day. Our local data catalog was fundamentally built around the principles of MapReduce rather than the principles of a database – it was created from scratch every day by hundreds of machines all running in parallel with our pipeline and all our algorithms and models.

We did have a mechanism of storing past corrections we had received – these were fed into the pipeline as another data feed. But by basically recreating the catalog from scratch every night, we could fully leverage any improvements in our machine learning models. So as an example, if an improvement was made to our match function that dramatically improved our duplicate rate, we could instantly benefit the next day when we reran all existing data through our pipeline with the new match function. We also reduced the risk of a bug introducing a long-lived corruption or degradation to the catalog since we were throwing it away every night anyway. A bad model or bug would at most impact the catalog for a couple of days until we caught it and fixed the model or bug. We also had the ability to roll back to a previous catalog easily if there was a bug in a model or in our pipeline.

One-Off Corrections   When we had a conflation error in production (e.g., duplicate businesses that needed to be merged), a manual correction system allowed us to quickly fix those issues. But the storing of past corrections proved to be a difficult system to manage. Manual corrections made to the conflation system tended to cause major quality issues as those corrections aged. As the conflation system got better and as input data changed, the manual corrections often got in the way, and corrections went from correcting the data to causing data errors. A lot of time was spent trying to figure out how to age out or remove corrections that were bad. Ultimately, human judges were used to audit on a regular basis which corrections were still correct and which were no longer valid.

As hoped, we did indeed achieve great gains by using the approach of rebuilding the catalog from scratch each night. As we improved our conflation models, extraction models, attribute selection models, and so forth, we could instantly see large improvements across the catalog when these changes were deployed. And we avoided the complexity of trying to track these changes in a database and reduced our reliance on human editors for data.

There were some significant drawbacks however to this approach. First of all, we were shortsighted in building our data pipeline in a fundamental "build the whole world" approach rather than supporting incrementality in a meaningful way in our data pipeline. Even though on a particular day, for example, Yelp might only be giving us a couple thousand local business entities that were different than the day before, our pipeline still ingested, processed, conflated, and published everything regardless of whether the source data for a particular entity had changed in any way. As the number of data feeds we got increased, our pipeline became less performant. As the number of models we wrote increased, our pipeline further degraded in performance until it was taking multiple days to build a new catalog. This reduced the freshness of our local data catalog.

A second significant drawback we had related to what we called "ID churn." One thing that you get from a database system is a fundamentally stable ID. In a large MapReduce system like ours where data is re-clustered every day, those clusters can shift from day to day as they are fundamentally linked to the match score which can also shift as the match function is improved – ID stability became an issue for us. We incorporated several ID stabilization techniques in the "ID enrichment" stage in Figure 2-4, but we were never able to achieve the level of ID stability that a database system allows.

Having a system that "built the world from scratch" every day ended up being a competitive advantage for us – enabling us to change our catalog, algorithms, and models much more quickly than if we had adopted a database-style architecture.

Tests and Monitoring to Enable Change

Since we had a "build the world from scratch" system, we had to be innovative about how we detected that today's catalog wasn't catastrophically different than yesterday's catalog. We did that in three ways: monitoring incremental change in the catalog, monitoring "sentinel" entities, and computing a quick quality score for our data.

Monitoring Incremental Change: The Data DRI

We first developed an extensive and configurable system to monitor and alert on significant changes in our data from run to run. For example, we could monitor the number of phone numbers we published on Monday and compare it to the number of phone numbers we published on Tuesday when we rebuilt the catalog from scratch. Thresholds could be configured in the system to determine over time how much of a change in the number of phone numbers was alarming and worth investigating. Since all our data sources were reacquired every day and we used machine learning models to pick the best phone number from many different sources, it was not unreasonable to see several thousand phone numbers in the catalog change from day to day. So we established baseline thresholds for acceptable change in the catalog through observation for several weeks and then set alarms that would go off when those numbers changed beyond those thresholds.

When one of those monitors would go off – for example, the "an alarming number of phone numbers changed today" – a data DRI would begin to investigate. DRI stands for "Designated Responsible Individual," and each week a team member would be assigned to be the DRI. The DRI had a set of tools at his or her disposal to investigate changes in the data, and after some investigation of the changes that had been made to the system since the last run of the pipeline and some manual verification of some of the phone numbers that changed, the data DRI would decide to either continue the publish of the catalog or trigger an additional investigation by the team.

With this system, it was very important to invest in tools to speed the ability for the data DRI to make a "publish/no publish" decision for the catalog. Also, these alerts could trigger at early stages of the run of the creation of the catalog – so while the data DRI was investigating whether to let the run continue, precious time was ticking that delayed the publish of our next set of updates to the catalog. So, in addition to making significant investments in tooling so the DRI could explore what had gone wrong with the pipeline, we also made significant investments in tuning alerts and reducing the number of false alarms in the system.

Sentinel Entities

A second embarrassing occurrence that happened more times than we liked was when a very important entity suddenly disappeared from the catalog on a new run – for example, we had several occasions where large casinos in Las Vegas mysteriously vanished from the catalog, and there were no Vegas magicians involved.

To solve this problem, we created a list of hundreds of "too important to ever lose" entities that we would explicitly make sure were in the catalog every day. Dropping one of these sentinel entities (also sometimes called "hero" entities) was serious enough that we would always stop the publish of the catalog and investigate what had happened.

Daily Judged Metric

A third way we monitored our "built from scratch every day" system was to do a daily human-judged metric. Doing human-judged metrics is expensive and takes time, so we did a lot of work to reduce the actual number of judgments we had to make when a new catalog was published, taking advantage of the very small rate of change in our problem space of local data. Big shifts in this metric were usually due to engineering errors, not real shifts in the data.

Our daily metric was based on a set of about 1500 entities that were sampled at random from the catalog at the start of a 6-month period based on popularity of entities. So more frequently shown entities were more likely to be in the set of 1500 than less frequently shown entities. At the start of the 6-month period, the entities were exhaustively judged for accuracy of name, phone, address, category, latitude and longitude, web site, and so on. Then, when each new catalog was published, we used a variety of techniques to try to find those 1500 entities in the new catalog (since same ID was not already ensured, we would try by ID first but then fall back to other matching methods). We would then calculate what attributes changed between the old publish and the current publish and send those attributes immediately to judges to judge, and then a recalculated metric assessing the overall accuracy of those changes would come back usually within an hour.

This metric was another great tool for detecting unwanted change in our database and also provided great insights into how our algorithms and models were affecting 1500 of our key entities. We were able to build a variety of tools to track these 1500 entities as they were created each time our pipeline ran to better understand the performance of our system.

Metrics Can Sometimes Double as Integration Tests   If you can develop metrics that can be recomputed frequently – ideally daily or even better on each build and run of the system – these can be used as integration tests as well. We had dedicated integration tests, but often our daily metric would catch regressions in our system that the integration tests wouldn't.

Testing Features

The basic building blocks of many machine learning projects are the features that are extracted from the data. In our system, we had two types of features we implemented extensively. The first type was an entity attribute similarity feature that took two entities as input and computed a similarity score. The second type was an entity additive data feature that took a single entity and some corpus of additional information or algorithms which would add additional data to the entity through inference.

For example, some entity attribute similarity features that might be extracted from a local business entity that could be used for matching included a feature that calculates a similarity score between two attributes of businesses ingested by the system. Possible attributes that could be compared included business names, locations, categories, phone numbers, and so on. Possible algorithms to use to compute similarity scores could include the Euclidean distance between two points or an edit distance between two names such as a Levenshtein distance.

Some examples of entity additive data features included comparing the phone number of an entity against known catalogs of active phone numbers and adding an attribute to the entity called "KnownToBeGoodNumber." Another example would be to do spelling corrections on a name to generate a new attribute on an entity called

"SpellCheckedName" or using an address validation component to add an attribute called "ValidatedAddress." These additional entity attributes could also be combined with entity attribute similarity features once created.

The set of features we implemented was continually growing as new inference techniques and algorithms were explored and grew over time to include thousands of features. To enable change in the set of features, two techniques were useful. First, we ensured that all new features had good unit testing coverage. Second, we built many of the building blocks that were commonly used in feature development into a shared common library that could be used by feature writers to quickly develop and create new features.

For a while, there was debate on the team about how much unit testing we needed to have. Because features are ultimately composed into larger inferences on the data and because the quality of the generated data is measured by metrics, bugs in features often are detectable because metrics decrease – for example, the overall quality score for phone number accuracy in the catalog would usually reflect the impact of adding a bug in the "KnownToBeGoodNumber" feature. Ultimately, metrics regressions are one useful way to test your system, but we found we benefited by detecting failure as early as possible in the system – and a robust set of unit tests as well as high-quality shared common libraries assured that we could rapidly change features and find failures as early as possible as well as closer to the point of failure, thereby improving our rate of additional feature development.

Testing Learned Models

All the features that we developed were then leveraged by additional machine-learned models – typically boosted decision trees – that learned which of the hundreds of features were important to the data problem at hand and which were not. This process is known as "feature selection" in machine learning.

For example, one model we had predicted whether a given business entity in our catalog was closed. It consumed a number of features and was trained on labeled data of entities that were verified in the real world to be closed. Once the model was trained, it determined the importance of each feature to whether a business was closed or not. Table 2-2 shows some of the most important features as determined by the learned model. Provider A looks like it has pretty accurate closed data when it is available for a business. Also, the number of records in a cluster that concur on the entity being closed is an important feature (again think of a cluster as being what is shown in Table 2-1).

Table 2-2. *Importance of various features as determined by the machine-learned closed model*

Feature	Relative Importance
Provider A says the business was moved or renovated	4.5
Ratio of closed entities in the cluster	4.3
Provider A says the business is closed	2.5
Number of entities in the cluster	1.4
Provider B has an entity in the cluster	1.1
Provider C says the business is closed	1.0
Provider D has an entity in the cluster	.9

Labeled Training Data

Another critical thing to get right in the system is the quality and freshness of labeled training data. We utilized many labels in our system. For example, in our conflation system, we trained our match function with thousands of examples of entities that match and don't match.

Here, the challenges were to have a wide variety of fresh labeled training examples, provide clear judgment guidelines about which two entities were a match and which ones were not, and monitor the quality of judgments being provided by the judgment team.

Having high-quality labeled training data is really the secret sauce to having a strong product. We put a lot of emphasis on judge management and education as well as auditing of the training data. The ideal setup for us was we would hire judges who would sit in proximity to the development team – ideally on the same floor. The development team would work with the judges to write very detailed judgment guides for labels. For example, our judgment guide for primary web site for a business was 48 pages long and entirely created by the development team to ensure that the labels we were getting for web sites (we had six) were correct. The 48-page long guide didn't happen immediately – initially, we thought judging primary web site correctness would be easier. But over time the judgment team in the process of judging web sites would discover new scenarios, labels required would change, the development team would discover they needed more data for one feature or another, and so the guidance to the judges would change. Judges would also often come to the development team for judgment advice on a particularly tricky issue which would then lead to new features in the system or clearer judgment guidelines.

Learn from Judges   Judges look at hundreds of examples of data every day and end up understanding the data in some cases better than the development team themselves. Keep the judges close and meet with them frequently. Write down all the strange cases they come across in judgment guidelines. Judges can often inspire the development team to create new features as they describe on a regular basis to the team the strange real-world cases they come across. It is a good practice to have a weekly meeting between the development team and judges to hear what is really going on in the data and the types of issues judges are encountering.

As an example, initially in our labeling, we asked judges to answer the question "Is the web site for the business – say `http://www.walmart.com` – the official web site?" The initial label was Yes, Not Sure, or No. Initially, the Walmart.com site was acceptable. But over time, we realized that users wanted store-specific URLs so `https://www.walmart.com/store/3098/bellevue-wa` was preferable to the root Walmart.com domain, and so we added the label "No – More Appropriate Primary Site Exists" to give to a web site like "`www.walmart.com`".

Other techniques we would use to ensure quality labels included having more than one judge do the same labeling task for the same entity and then detecting when multiple judges disagreed on the label for a particular entity. We would regularly review the performance of judges who were providing incorrect labels and work to further train those judges and update the judgment guidelines to ensure that labeling challenges were clearly documented and explained so future labels were correct.

We strongly advise that the development team stays fully engaged with the judges providing labels and invests in creating better and better judgments. In particular, we did not use mechanical turk-style judgments unless they were extremely simple judgments – it is very hard to get the quality of label you want from crowdsourcing when your judgment guidelines for just judging web site accuracy are 48 pages long.

Responding to Customer DSAT

In Bing, we used the phrase "customer DSAT" to refer to a specific issue in our data or user experience that caused customer dissatisfaction – "**DisSAT**isfaction." For example, a common customer DSAT might consist of a business that we listed as being open from 9 to 5 on weekends but it was closed on Sundays. An even worse DSAT would be a business that we listed as being open from 9 to 5 on weekends but it had gone out of business or moved to a different address.

One of the biggest challenges with customer DSAT was deciding whether it was a larger problem that required a big team investment and new strategies to solve or whether we should just "hot-fix" the specific issue at hand and continue our current work. The best example of this dilemma was when the team would get a much dreaded "vice president (VP)"–level DSAT. One example on the team was when a vice president went to a nearby city in Vancouver and was going to go to a dress shop. The VP searched for the dress shop on Bing – it showed as being open. The VP then walked several blocks to the dress shop, but the dress shop was found to be out of business.

In our area – maps and local data – having a DSAT like this has a real-world consequence. Someone, in this case our vice president, took a possibly pleasant walk that then became unpleasant when it was clear it was a walk to nowhere. And then the questions the team was asked were, first, what did we do wrong to provide the bad business and, second, does it represent a new class of issues or is it an outlier that there really was no way that we could have detected given the data that we knew about that business.

In the aggregate, our "closed precision" as we called it was pretty high – it was in the high 90s that we would get closed right. Our competitors had similarly high numbers. But in a catalog of 10 million local entities with a closed precision of say 98%, you would still get 200,000 entities which present the wrong closed value. If you move to 99% precision, you still have 100,000 entities with the wrong closed value. It is basically impossible to achieve 100% precision on something as fluid and changing as whether a business has gone out of business or not – especially when there is little incentive for the person who has a failed business to let the world know about it.

In the case of the out-of-business dress shop, when we looked at everything we knew in the system about the dress shop, there was no signal we had that could tell us that the dress shop was closed. All of the various data providers we had that provided us with information about the dress shop all agreed that the dress shop was open. Our competitor also had the dress shop open. So to fix this particular DSAT, we just manually corrected the entity with the new piece of knowledge our vice president had gained after a multi-block walk to the dress shop.

Identifying Classes of DSATs

Sometimes though, patterns do emerge in customer DSATs that can lead to the generation of a "class" of DSATs that can be solved together rather than as a one-off fix. These patterns are seen as the team diligently postmortems each customer DSAT to determine what happened and if there was any data in the system that could have prevented the DSAT.

When we first launched the Bing application for iPhone and Android, we had a feature that listed all the restaurants in an area close to the users. Although we always had that data in our catalog, typically the way Bing was presenting restaurants was the top ten most popular restaurants in a several-block area. In contrast, the way the mobile app worked though was that it would present all the known restaurants in a particular area, and it suddenly became clear that the application was displaying more of our "tail" or less popular entities which had lower quality than our "head" or more popular entities. The mobile application was basically exposing a data issue that previous presentations of our catalog had hidden through just not having an effective mechanism to present more than ten or so results.

This new application created a data crisis where we began to realize we had real issues in our long tail of less popular entities. This had never shown up previously on our data metrics because we sampled the entities we judged based on popularity – how often users clicked on the entities the search results – we called this an "impression-weighted" metric. Basically, every time an entity got clicked on, it got put into a "bag" that we would draw out of when generating our measurement sets to measure the quality of our data catalog. So a popular restaurant like, say, *Thomas Keller's French Laundry* in Yountville, California, is much more likely to be measured than a less popular restaurant in the same area or, even worse, a closed restaurant that is still somewhere in the data catalog showing as open. We called these "tail" entities that were clearly bad, old, closed entities "junk entities."

As we dove into this new "junk entity" crisis, it became clear that there were some specific signals in our data that we could leverage to reduce the number of junk entities. Specifically, we found these junk entities often came from data providers that had data originating from scanning of phone books and whose corpuses were more geared toward generating bulk mailings to people where having a bad entity at worst resulted in an envelope being returned to the sender – not a customer walking out to a business that no longer existed.

We were then able to focus in on this class of DSATs, and we devised a number of projects and model changes to remove these junk entities from our catalog. As one example, for restaurants, we trained models that weighted higher a signal that a business was opened or closed from a provider whose data strength was restaurants rather than phone book scanning. We also generated a new set of metrics that were not "impression weighted" where any entity in the catalog had an equal chance of being picked for measurement as every other entity. This then generated for us a different set of quality metrics which more clearly measured our junk entity problem and the number of entities in our overall catalog that were bad.

Junk Pizza Hut Entities   A fun example of the problem of junk entities was when a partner team exhaustively listed all the Pizza Huts we had records for in a particular metropolitan area. Many of them were no longer open because they had gone out of business. This led the authors to discover the particularly interesting world of what has happened to all the Red Roof–style Pizza Hut buildings that were built en masse in the 1970s and 1980s. The book *Pizza Hunt* (www.pizzahunting.com) shows how many of these out-of-business Pizza Hut buildings are still being used today as everything from flower shops to doctor offices.

Regular Self-Evaluation: Data Wallow and Quality Reviews

Sometimes a data crisis is evident to everyone as was the case with the junk entities we were seeing in our new mobile treatment of search results. Other times, it is not as evident as the team is busy building new features and a data crisis is not properly identified.

Two techniques we found very useful to ensure that we were regularly understanding our own data problems were the "data wallow" and the "quality review."

The data wallow is where a subset of the team – perhaps four to six people – would meet for an hour and look very deeply at some slice of the data to understand a particular data problem. For example, as we worked to identify ways we could solve our junk entity problem in our data corpus, we spent a lot of time looking at long tail entities, exploring the provenance of those entities and whether additional signals were available on the Web or from other providers or sources to classify a particular entity as being junk or not. Since junk entities were primarily entities that were valid at some time in the past, one source we discovered through data wallows was that we found that for some of our junk entities, we could find reviews on the Web that said things like "this was a great restaurant, too bad that it has closed." So we invested in web extraction and NLP to detect these signals in reviews we could find for particular entities. Data wallows are described in more detail in Chapter 6: Effective Communication.

Data wallows would be combined with another regular practice that in Bing was called the "Search Quality Review." In this review meeting which typically happened every couple of months, we would aggregate together all of the DSAT that was reported in the last couple of months and try to identify if there were any patterns of DSAT that indicated we had a data crisis as shown in Figure 2-7.

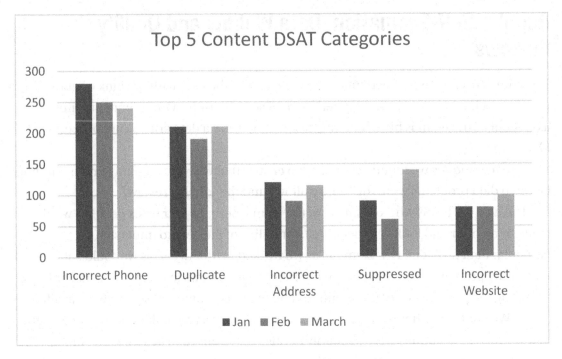

Figure 2-7. *Example of top aggregated DSATs compared over three months – numbers fictionalized*

We also used these regular quality review meetings to look at a scorecard of our top-line metrics. Figure 2-8 shows an example report of which numbers had improved since the last quality review and which had gotten worse since the last review. We also compared these with measurements of our key competitors to see where we were relative to key competitors.

US Query-Weighted Catalog Quality

Attribute	Bing (P/R) Current		Bing (P/R) Six Months Ago		Google (P/R) Current		Google (P/R) Six Months Ago	
Name	97.7	100.0	97.2	100.0	98.8	100.0	98.6	100.0
Phone	91.0	99.0	92.5	99.3	97.2	99.1	97.5	98.4
Address	93.1	100.0	94.4	100.0	97.0	100.0	97.0	100.0
URL	92.6	87.2	94.2	86.7	97.9	95.0	96.3	94.5
Open Precision	95.3		96.1		99.1		99.0	
Q	88.9		91.2		96.0		95.9	
Duplicate Rate	5.3		3.4		1.0		1.6	

Figure 2-8. *Example of a local data scorecard – numbers fictionalized. Q is a composite weighted score calculated from the other scores.*

Given these as our guiding principles of what needed to be focused on, we would create a prioritized list of techniques and projects to improve key DSAT areas and drive up metrics where we were either decreasing in quality or where a competitor was far ahead of us as shown in Figure 2-9.

Techniques to mitigate top DSAT	Website	Duplicates	Open/Closed	Hours	Location	Name	Phone	Menu URL
SMILE	X	X	X	X	X	X	X	X
2nd Conflation		X						
Containment		X						
Enterprises		X						
Departments		X						
Closed Classifier			X					
Moved Classifier			X					
Definitive Feeds		X						
Bad Address Classifier		X						
Publish Confidence Score		X						
Bad Link Filtering	X							X
ML-URL Merge								X
ML-Menu Merge	X							
Hours Extraction								
Machine Assisted Curation	X	X	X	X	X	X	X	X
ML-Name Merge						X		
ML-Phone Merge							X	

Figure 2-9. *Techniques to address key DSATs after a quality review*

Elements of a Successful Quality Review   Follow these steps to have a successful quality review:

1. Look at and root cause as many DSATs as possible since your last review.

2. Categorize the DSATs and rank the DSATs in order of concern.

3. Determine your key top-level metrics and create a scorecard.

4. Look at both your ranked DSATs and top-level metrics and create a prioritized list of areas of focus.

5. Use what you have learned from doing root cause analysis of DSATs to propose areas to improve and prioritize that work for the upcoming time period.

6. Repeat at regular intervals.

Measuring the Competition

One advantage we had in our space is we had very strong competitors that we could measure the quality of regularly in addition to our own quality. This often helped us to identify areas where a competitor was doing better than us and gave us confidence and motivation that we could improve a given area of the product. Although we would regularly measure a competitor, we ensured that competitor data did not get into our own catalog. The "ground truth" of whether a particular business was correct or not was typically determined through calling the business directly. If that was not possible, we would rely on the primary web site of the business or data provided to us directly from the business owner. We did not consider any competitor data to be ground truth.

Through our measurement, we learned several things. First, we learned to be skeptical of our own measurements and to adjust and refine our measurements over time. As an example, we developed a metric called "Q" for quality. This metric was designed to be a composite metric that represented the overall quality of our catalog. As a composite metric, it was easy to just communicate it as a short hand for several other metrics including things like name accuracy, phone accuracy, phone coverage, closed accuracy, and so on. Q was a great goal-setting metric for us – we would shoot to gain several points of Q against a competitor over the course of 6 months, for example.

But it became clear as the Q gap became smaller between us and our competitors that Q wasn't correctly capturing the quality gaps between us and our competitors. We measured how well we were capturing the quality gap by also asking end users if they preferred a "left" or "right" result where one side would be provided by us and the other side by a competitor.

We learned several things in these user "preference polls." First, we observed that there is a brand inflation that occurs with certain brands. For example, when presented with a left-right comparison between our catalog and Yelp, people tended to think Yelp was correct even when the ground truth said that we were correct. People had grown to trust Yelp especially for restaurant data.

Another thing we learned was that our composite "Q" metric was too simplistic over time to truly reflect the gap in quality between us and competitors. For example, people valued Yelp data over our data because of the number and quality of reviews provided of restaurants. We noted that people valued Google data over our data because the quality of what we called "rooftop lat longs" or the point on the map where the pin is dropped was more accurate than our pins were. Our pins were basically initially generated based on geocoding the address of the business and tended to be good for driving directions. But when people did searches on maps, they wanted that pin to be squarely on the building or the entrance to the building on the map rather than street side. So we added "roof top lat long" to our composite Q metric. This increased the gap in the measurement between us and Google, but more accurately represented our true state relative to our competitor and motivated us to find new data sources and techniques to improve rooftop lat longs.

Be Skeptical of Your Own Metrics   especially if the metrics say you are winning in a particular area over a competitor, but real customers don't feel that way. It is important to continue to evolve your metric and dial it in and refine it over time so that the metric motivates and informs the team about where the current gap is with a competitor and generates the forward momentum to continue to close the gap.

It is also very important to keep these metrics internal, and especially don't share them with marketing or people outside the team who don't understand that the metrics are an evolving way to measure – otherwise, you may have someone saying how much better you are than a competitor where there isn't really a true advantage. Help people understand the margin of error on a metric. For example, in Figure 2-8, the slide says that Bing's name accuracy is 97.7 and Google's name accuracy is 98.8. We learned over time that reporting numbers with this level of specificity – to the first decimal point – is misleading because the margin of error on these numbers would vary per measurement by as much as two points. So it was within the margin of error that Google's name accuracy was actually 99% and our name accuracy was 95%. Also, as previously discussed in this chapter, our metrics were impression weighted meaning that more popular entities were more likely to be measured than tail entities. When we discovered our "junk entity" problem, our name accuracy was measured to be much lower when we began to measure the quality of the tail. Ultimately, we began reporting out both impression-weighted results and un-weighted results to more accurately reflect the quality of our catalog.

Conclusion

In Chapter 2, "Changing Requirements", we have discussed how you can build and develop systems anticipating change; tests, monitoring, and measurement to anticipate and measure change; and strategies for responding to DSATs and measured problems in your system.

In Chapter 3, "Continuous Delivery", we will discuss how to ensure that the team is continuously building, delivering, and verifying software and data.

CHAPTER 3

Continuous Delivery

*Deliver **working software frequently**, from a couple of weeks to a couple of months, with a preference to the shorter timescale.*

—agilemanifesto.org/principles

The Agile principle for this chapter describes how frequently a team delivers working software. We extend this to include frequently delivering data as much of what a data engineering team delivers is high-quality data. We further extend the principle by replacing the word "frequent" with "continuous" – we believe it is important to deliver working software and data continuously. Our rewrite of this principle would be "Deliver working software and accurate data continuously." On any given day, the team should have software and data available that effectively represents the incremental work done on the previous day.

How do you make the move from frequent delivery to continuous delivery? We will consider how this is done with software and then with data. With software, modern engineering teams practice techniques called "continuous integration" and "continuous deployment." Continuous integration is enabled by developers verifying their code changes before submitting them and then running automated systems that fetch the latest code changes made by developers, build those changes, run tests on those changes, and finally deploy those changes – typically to a series of environments over time that include more and more customers.

Verifying Code Changes

For an automated system to continuously build the product, there must be a mechanism to verify code changes before they are checked into the code base. This part of the system must be carefully designed and respected by all developers to ensure that only

E. Carter and M. Hurst, *Agile Machine Learning*, https://doi.org/10.1007/978-1-4842-5107-2_3

high-quality code changes are submitted. Some of the things that are typically done to ensure that code changes are of the highest quality include the following:

- Coding standards and preferred patterns are well documented and understood on the team and, where possible, checked by automated tools that a developer can run on their local machine before submitting code changes.

- On a local machine, the developer can build the system they are working on and all other dependent or potentially impacted systems to verify no build breaks have occurred.

- The developer can quickly run and verify a modified system either on a local machine or in an online developer environment that is representative of the final environment where the code will run in.

- The developer can quickly run some set of unit tests that verify assumptions made in developing the system – these unit tests are extensive enough that they cover a high percentage of the system (we usually shoot for 80% coverage), and the tests are written in a way that they can run extremely quickly – typically by using patterns to mock parts of the system that are expensive to start up.

- New code changes are checked in with appropriate unit tests or integration tests to ensure that test coverage remains high.

- A code review process is followed where any changes to the code are inspected and double-checked by another developer on the team – typically one who is more experienced in the system where possible. Any suggested changes as part of that review are made and reverified.

- The developer can quickly run a small set of integration tests that exercise the system end to end – these tests start up the whole system in either a local machine environment or an online development environment and test critical functionality of the system that must always be working.

- The developer syncs to the latest changes in the source code base
  and re-merges changes if some other developer has checked in
  conflicting changes to the same area of the system being worked on.
  Whenever a change happens as part of this process, the change is
  rebuilt and reverified by running tests again.

- Once all has been completed, the developer submits final code changes.

Keeping an Eye on the inner loop   Note that for these steps, every effort should
be made to keep the number of tests and verifications that are run by a developer
as fast as possible. Sometimes this stage of verification is called the developer
"inner loop," and we will have more to say about measuring the health of this
inner loop in Chapter 5: Motivated Individuals. More extensive tests that are longer
running can be placed later in the continuous integration "pipe" which is called the
developer "outer loop." If issues are found in a check-in made by a developer later
in the pipe, the change can be automatically rejected by the system and sent back
to have the developer figure out what went wrong. But having a change rejected
later by the outer loop can also be costly, so a balance must be struck between
finding the problem now in the inner loop before the initial code submission goes to
the continuous integration system and finding the problem later in the outer loop.

The Continuous Integration System

Once a verified change is checked in, the continuous integration system takes over.
Systems like Azure DevOps or the Jenkins build system can automate an integration
system for code changes. These systems typically watch for new check-ins to verify and
run a similar set of verification steps to what the developer did locally but now running
on all build targets and devices the team cares about while running all unit tests and
integration tests the team has available. If any step fails, the check-in is rejected by the
system, and the developer is notified along with logs so an investigation can be made into
what went wrong. Typical continuous integration systems run steps like the following:

- Queues an entire clean rebuild of all the code in the team's repository
  targeting all the build targets that the team cares about. For example,
  the continuous integration system might build code to target AMD-

64 processors in both debug and release flavors as well as code that targets devices like iPhone or Android devices.

- Runs a more extensive set of unit tests on all the build targets.

- Deploys the system to online development environments that simulate the production environments and runs a more extensive set of integration tests within that environment.

- Deploys compiled code to devices and runs tests on those devices (if applicable).

Sometimes this stage becomes too heavyweight, and verification takes too long – a good rule of thumb is this stage of verification should take around an hour. If this verification length becomes too long, then the number of tests run on an individual change should be reduced to bring the verification length down. Tests that now no longer run on each individual change should still be run but at a lower frequency on batches of changes. For example, all the changes submitted in the past 12 hours can be batched together, and the tests that were moved out of the individual change verification process can be run on the larger batch of changes. If a failure is detected in this stage, more work is required to determine which change in the batch caused the failure, but it is typically worth the improved speed of verifying individual changes.

Continuous Deployment Systems

If all these verification steps are successful in a continuous integration system, the final step of such a system is sometimes called a continuous deployment system. Continuous deployment systems do what you would expect – they take a set of changes that have been verified and deploy them automatically. Typically, deployment doesn't go straight to all your users as there is still significant risk that some additional issue may still be in a code change that even the best unit testing and integration testing won't catch. So continuous deployment systems typically deploy sequentially over time to a series of deployment environments often called "rings" that represent a continuously growing set of users.

Typically, the first environment where a newly verified code change is deployed to is called "Ring 0" or the internal ring as shown in Figure 3-1. This ring is the deployment environment that is used daily by the development team and maybe additionally by a set of early adopters of the system. Each ring is built in a way that if a change is deployed to that ring that causes live site issues to users of that ring, the change can be quickly rolled back and the ring can be reverted to its prior state. As you might expect, the devil is in the details, and software systems under continuous deployment must be carefully designed to ensure that they can be rolled back to a prior state. As an example, consider a change that is deployed to an environment that changes the schema of a database being used by the system. If this change fails in a ring, there must be a mechanism to not only revert the software but bring the database back to the previous schema. There are many techniques to make systems function well under continuous deployment that are beyond the scope of this book.

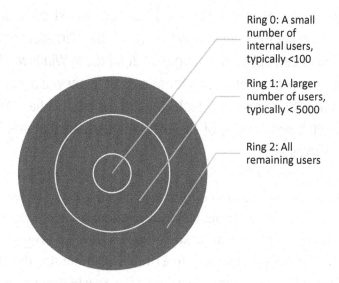

Ring 0: A small number of internal users, typically <100

Ring 1: A larger number of users, typically < 5000

Ring 2: All remaining users

Figure 3-1. *Deployment rings*

If the team and early adopters don't encounter any issues in the internal ring, the continuous deployment system then pushes the same set of changes to a larger environment that impacts more users. Sometimes deploying to the next ring is automatic after a certain amount of elapsed time; sometimes teams control this decision manually. If the change is successful in the next ring, it is then deployed to additional users until the change is visible to all users of the system. Part of the creation of a continuous deployment system is deciding how many rings of users you want to support and which users will be in each ring. We typically see systems that consist of an internal ring with mainly team members participating, a company ring which includes everyone in the company, then an early adopter ring which begins to include those who have opted in to use the latest and greatest changes in the system, and finally an "everyone" ring which includes all users of the system.

Eating Your Own Dogfood   There is a rich tradition at Microsoft of what is called "dogfooding" which refers to using the product that is still in development. This came from an email a Microsoft manager sent with the title "Eating our own Dogfood" which was advocating increasing internal usage of a software product that was in development. Microsoft Windows has steadily extended dogfooding to even participants outside the company with a very active "Insider" testing program where external users can opt in to testing early features in Windows. These types of programs have been very successful for Microsoft as they allow both early feedback on features that previously took years to get into the hands of real users and early testing on a wide variety of system configurations that aren't always available within Microsoft.

When something goes wrong in a continuously deployed system, we mentioned that rollback is the ideal solution. Other techniques are sometimes used as well for more flexibility in fixing a broken feature. Sometimes, a "hot-fix" is employed – these are one-off fixes that are deployed directly to a ring to fix something that is broken in a ring. Ideally, a hot-fix can be a change to a config file, so whenever possible, features are implemented in a way that they can be modified easily with config files. But config file hot-fixes, though easier than changing binaries deployed with the system, can be just as dangerous and must be well tested. Hot-fixes can also be made by applying some small number of code changes, recompiling the impacted binaries, and patching the ring with those newly updated binaries. Hot-fixes should be run through the same set of automated verification steps in the continuous integration system if possible.

Another technique that is used to ensure a poorly behaving feature can be disabled is making sure a newly shipping feature is "flight-able" and can be turned on and off easily through configuration settings. Flighting is actually very similar to the system that enables rings to function – it is the ability to make a new feature available to a subset of your users and compare their experience through metrics gathering to the users who don't have that feature available. A small percentage of your users can be on a flight to try the new search experience your team has just developed, while the rest of your users have the old search experience. You can then monitor things like stability of the system for the new search experience users vs. the old search experience users. If you see significantly poorer numbers for the new search experience, you can then immediately turn off the new search experience for all users while you investigate through looking at logging what was going wrong in the new system. We talk more about this in Chapter 7: Monitoring.

Verifying Data Changes

All previously discussed techniques for continuously integrating and deploying code changes can also be used to continuously integrate and deploy data changes. These techniques include the following:

- Developers on data engineering teams are typically writing code that is changing data. Just as code that modifies data must be reviewed, the data itself should also be reviewed when it is modified by code. It is usually not possible to exhaustively review all of the data changes that happen in a system when code that modifies data is changed, but a sample of the changes made to data by the code can be taken and examined carefully to ensure that no unexpected changes are made.

- A specific sampling approach should be taken when examining changed data. Data that did not change at all should be sampled from to verify that the "non-change" is correct. Data that was deleted from the system should be examined to make sure the removal of that data is correct. Data that was added to the system should be examined to make sure the addition of that data is correct. Data that was only slightly modified should be sampled and examined. And data that was more severely modified should also be sampled and examined.

- Data standards and preferred patterns for data are well documented and understood on the team and, where possible, checked by automated tools that a developer can run on a local machine before submitting the data change or code that causes data changes.

- If the amount of data to be reprocessed due to the change is small enough that it can be run on a local machine or in an online developer environment that is representative of the final environment, all data impacted by the change should be regenerated and verified. In many big data projects, this isn't possible, so instead some representative slice of the data should be regenerated and verified. This slice should be sampled as widely as possible so that the probability of the slice containing all the types of patterns seen in the larger data set is high.

- A data review process is followed where any changes to the data are inspected and double-checked by another developer on the team – typically one who is more experienced in the system where possible. Any suggested changes as part of that review are made and reverified.

In addition to just manually inspecting sampled portions of the data, we talked in Chapter 2: Changing Requirements about several techniques that can be used to inspect the data in an automated way. These techniques can be used on developer machines, as part of the continuous integration system, and as part of the continuous deployment and ring systems:

- Data "gates" can measure frequencies of patterns in the data both before and after a change is made to examine whether an extreme change has taken place to the data that may be a bug. These gates are determined by the team through long experience gained from manually inspecting and learning about the data they work with. For example, in Bing's local data team, we observed that shifts of greater than 1% in things like phone numbers in our corpus were much more likely to be caused by a bug in our system than by real-world variance and change to actual phone numbers coming from data providers. Many such rules of thumb can be learned about the data your team works with and can be coded into a data gate system that can be run after every change to your data.

- Sentinel entities are entities in your data that you know represent absolute truth. These entities can be monitored and reverified after every change to your system to ensure they remain constant.

- Human-judged metrics can be recalculated daily on your data to ensure that some known set of entities in your system is continually improving in quality.

Focus on the Changes   In Bing, we had a set of 1500 entities that was judged every day. Actually, we would only judge the changes made in that set of entities every day which usually represented only a small number of changes that could be looked at by human judges in under an hour. We could constantly monitor whether that set of entities was getting better or worse due to code changes made in our system. Specific judgments that were judged as worse could be looked at by developers to determine if they had introduced a bug in the system or if it was just a variance in a learned model.

Continuous Deployment of Data

Just as with code changes, data changes should be automatically verified by continuous integration systems and by continuous deployment systems as a data change rolls out to one ring, then another, and then another. Verification occurs via data gates, sentinel entities, and judged metrics computed as new data comes online in a new ring. Data changes are typically much more extensive than code changes, so it is much more difficult to "hot-fix" a data change – although the code change that caused the data change can sometimes be hot-fixed. Rollback systems for data are usually expensive especially when there is a lot of data involved.

In Bing, we maintained two systems of data storage, and one was amenable to roll back and the other was not. We were able to keep a smaller set of entities in a quicker to change and roll back system that was our "fast" store. We had a much larger set of entities that were kept in a "slow" store that was much slower to change and roll back. It turned out that by keeping about 2 million entities in our fast store, we were able to cover 80% of our page views – that is, 80% of local queries in the local data space were for the most popular 2 million local business entities. The other 20% of queries were for our longer tail of local business entities – about 18 million additional entities. So we kept

67

2 million entities in our fast store and 18 million in our slow store. This way, if we had a catastrophic data error, we could roll back the 2 million entities in minutes and fix 80% of our traffic, but the other 20% of our traffic would suffer until we could rebuild the slower 18 million–entity index which often took hours.

Deciding What to Ship

Just because you can deploy data and code continuously doesn't always mean you should. It is just as important to make decisions about what not to ship as what to ship. For features that are code only, features that don't impact data, you should instrument the feature with enough telemetry so you can determine whether users are actually using the feature and whether it actually meets the needs of the business. We will have more to say about telemetry in Chapter 7: Monitoring. Features should be implemented in a way that they can be flighted, and metrics with the feature on and off can be compared. It is also important to develop a feature in a way that it can be turned off and removed from the system easily. If telemetry indicates a feature is not being used or is not meeting business needs, it should be removed.

Just Kill It   You have just worked for several weeks on a feature, and you are now flighting it and finding that users aren't using it. Although you may try a couple of rounds of adjusting the feature to see if it takes off, you should not get too attached to a feature to kill it. It may seem like the several weeks of effort will be wasted when you don't ship the feature. But future maintenance nightmares can be far more expensive than the initial development cost of a feature. A simpler product that precisely meets customer needs is far better than a complex product that customers are only really using a small percentage of.

Decisions about which data changes to ship are trickier and usually need to be made earlier in the process. Here is what often happens in a project. A data scientist on the team has an improvement to a model that is ready to ship. How do you decide whether to ship it or not? There are many metrics that can be used to evaluate the "goodness" of a learned model. These are useful and should be measured and used as part of the evaluation of whether to ship a new model.

An additional practical way of evaluating a model is to look at the "wins" and "losses" of the model and decide whether the model is worth shipping based on that analysis.

When we would have an improved model that a data scientist would want to ship, we would sample from all the changes that the model made to our data and judge which ones were good changes, which were neutral changes, and which were bad changes. It was rare that a revised model would not introduce some small number of "bad" changes in addition to good ones. Sometimes the quantity of bad and good changes would be similar, but the severity of the bad changes would be greater. In this case, additional work would be done to see if the model could be further refined to remove enough of the bad changes while preserving enough of the good changes to make the model worth hipping.

New models could also introduce changes that weren't necessarily better but just different. It was important to not ship these models because the amount of random change they introduced just wasn't worth it. It would confuse our users to have us ship data changes where two correct attributes would be swapped. For example, businesses often have multiple phone numbers that all ring through to the same contact person. We would have a new model that would change the primary phone number, but this would just confuse the user since the number changed for the business but they still reached the same person at the front desk. Where possible, we would avoid shipping such models.

A Better Model May Be Too Costly   Although we would typically ship a model where the number and impact of good changes exceeded the number and impact of bad changes, we also considered an additional factor: the cost and complexity of the new model. Sometimes a new model would produce better data, but the cost of running it would be significantly higher. For example, the model might take multiple times longer to produce a result and thereby slow down the product or increase the amount of hardware required to efficiently run it. Sometimes a new model would produce better data, but the complexity of the new model meant it would be a maintenance nightmare to keep running and healthy. These less obvious aspects of an otherwise winning model should be considered and may sometimes lead to the rejection of a new model.

Conclusion

In Chapter 4, "Aligning with the Business", we will discuss the importance of and techniques for coordinating work with the business.

CHAPTER 4

Aligning with the Business

Business people and developers must work together *daily* throughout the project.

—*agilemanifesto.org/principles*

Ciao was a price comparison and product review web site that was founded in Germany and became the top shopping web site (shown in Figure 4-1) in many European markets attracting 19.6 million unique visitors per month in Europe by 2008. Microsoft purchased Ciao in 2008 to bolster its Bing search engine in Europe. Eric had the opportunity to work with the Ciao team as its new engineering manager after the Microsoft acquisition. What Eric observed was a development team that was extremely well aligned with the business – more aligned than he had observed most Microsoft teams to be in his prior work. There were many factors that allowed for this high degree of alignment. In this chapter, we will examine several of these factors that allowed Ciao to be successful.

© Eric Carter, Matthew Hurst 2019
E. Carter and M. Hurst, *Agile Machine Learning*, https://doi.org/10.1007/978-1-4842-5107-2_4

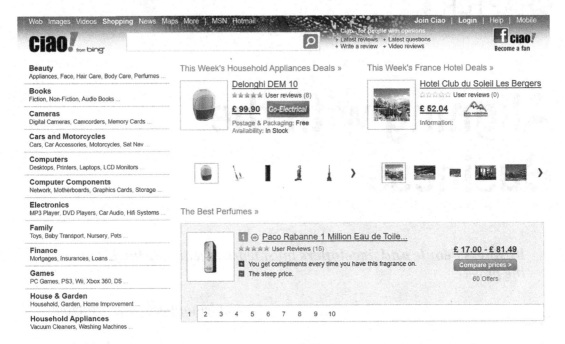

Figure 4-1. *The English language version of the Ciao web site in 2012*

The Importance of Daily

Perhaps the most shocking part for Eric of coming to work at Ciao after 10 years at Microsoft was that the business people and engineering people were all together in the same building and even often on the same floor. Software developers would go to the kitchen to microwave their lunch; and in that same kitchen, they would run into people who were selling the ads for the web site, people who were creating content for the web site, people who were marketing the web site, and people who were maintaining the servers hosting the web site. The people driving the business forward were literally found side by side with the people writing the code that the business was running on. In contrast, in Eric's time at Microsoft, he never met a marketing person, a person writing content for the product, or a person hosting the product.

This side by side-ness at Ciao didn't end at the lunch room. Business people having problems with the ad system on the web site would drop in to the daily scrums of the engineering team to explain their problem or to pitch an improvement. Software developers could walk over to the desk of a person having trouble using a content labeling system, they would observe what the challenge was, and often they would go back to their desk with a concrete idea for an improvement or fix.

For the purpose of this chapter, let's consider further who were the "customers" of the Ciao web site and who were the "business people." There were three types of customers. First, the largest population of customers were those that came to the web site to research something they wanted to buy and find the lowest price. The second customer group was the review community: people who came to the web site to leave a review which was further motivated by reputation and monetary rewards for helpful reviews. The last customer group were the merchants – third-party sellers that had a product to sell and wanted their price listed on the web site along with a link to their web site where the customer would ultimately buy the product they were shopping for.

The business people of Ciao were all the non-engineering staff that supported the needs of the three customer groups, kept the web site running, kept the money coming in, and kept the business moving forward. At Ciao, this included the following:

- Sales people who convinced merchants to list their prices on the web site and then supported merchants with tools to easily upload their price data on a regular basis

- Content people who edited and corrected the catalog of all the things one could buy

- Community people who moderated the review communities and made sure they were vibrant and thriving

- Marketing people who organized special campaigns with other companies – for example, they ran campaigns with certain car manufacturers to get Ciao members to test drive and review new car models coming out in a particular year

- Billing and finance people who ensured that the merchants that listed their prices on the web site were billed and paid for traffic redirected to their sites

- SEO (Search Engine Optimization) people who ensured that the site got steadily increasing traffic from popular search engines

- Operations people who ensured that the datacenter was healthy

In contrast, Eric's experience at Microsoft to that point had much more separation with business people who were playing similar roles for the products he worked on at Microsoft. At Microsoft, the "program manager" role interacts with business people and then translates their needs into requirements for the engineering team. Through the Program Management (PM) team, Eric would indirectly get guidance representing what many business people wanted to have happen in the Visual Studio products. This level of indirection was convenient for both business and development, but it also put the business people another degree of separation away from him as an engineering manager.

The Power of Shipping Daily   The jolt Eric experienced at Ciao of daily interactions with other people working in the business was further accelerated by the fact that he was moving from working on one of the world's largest Windows C++ applications (Visual Studio) which shipped only every couple of years to an application that was delivered via the Web and was updated every couple of days. The immediacy of being able to build a feature for the Web and ship it the next day to see if people liked the feature or not had startling implications for understanding the business and the customers of the business. In Visual Studio, a feature would be worked on for several years and finally shipped; and although customers were consulted with and demoed to frequently throughout the year, Visual Studio would often ship a feature that customers either didn't like or didn't use. On the Web though, this cycle shrunk dramatically where features could be shipped incrementally in a matter of weeks rather than years and the business impact of a feature could be immediately measured and quantified. In Bing, we went even further – changes were shipped daily thanks to the combination of flighting a new change to a small number of users and a system that could run extensive tests on any change to the system.

Advantages of Colocation

Let us list some of the advantages of having the business people on the same floor as the engineering people at Ciao and some of the ways that these advantages can be replicated even if building space and people location doesn't allow for true colocation.

One of the first advantages is that when a particular month is going well for the business, when sales are good, when traffic is up, there is a definite morale impact to have the engineering team drink in the good vibes and excitement among the business people. A successful month can be very motivational for an engineering team, especially when they see direct connection between the work they have done and the gains made by the business. It is also healthy in a month when things aren't going so well for developers to feel that as well. Often in those poor months, spontaneous conversations would happen between business people and engineering people that would lead to new feature ideas or improvements to the web site that could address flagging sales, dropping traffic, or customer complaints.

It is important to find ways to help the engineers feel some correlation between their work and the impact it has on the business. One good practice is to have a monthly business review to go over the key numbers of the business. Some of the numbers that can be reported including month-over-month changes include the following:

- Total visits, page views, and unique visitors to the site

- Organic visits vs. search engine–marketed visits

- Engagement (number of pages a user views before leaving the site)

- Time on site/time per visit

- Number of downloads of apps (if part of the business)

- Monthly active users (MAUs, unique users who have had some meaningful engagement with the site over the past 30 days)

- Monthly engaged users (MEUs, unique users who are much more engaged than the monthly active users – these represent your fans and typically are a small subset of your active users)

- Retention numbers (unique users who are continuously coming month after month and conversely unique users that you are losing month over month)

- Total revenue per month

- Operating metrics – at Ciao, this included how many merchants were listed on the site, how many offers (a unique merchant price for a particular product) were on the site, how many products were on the site, and how many reviews were on the site

- Availability metrics – how much downtime did the site experience over the past month and where

- Competitor comparison metrics – how much better/worse is the site than key competitors

- Comscore metrics (Comscore is a site that measures the popularity and traffic a web site is receiving)

- Any other metric that correlates well with business goals that are currently being driven

There are many ways to deliver these kinds of business metrics and content to developers. Some successful approaches we have seen include monthly business reviews conducted as optional meetings for the development team and business information sent out in monthly emails, available in always up-to-date self-serve dashboards, or presented in all-hands meetings. As development teams often prefer building products to reading mail or attending a longer meeting, we also recommend having a brief summary of business metrics as part of end-of-sprint demo meetings as well. It is well worth having several ways to communicate this information to the team to ensure it is received.

Physical Collocation Is Not Required   Of course, physical collocation is just not always possible. Chapter 6: Effective Communication talks about the Agile principle that face-to-face interaction is valued and provides some ideas for when true physical colocation isn't an option.

Business-Driven Scrum Teams

One thing that Ciao was very good at was mapping scrum teams directly to key business objectives. At a high level, the business goals were the following:

- Increase the amount of traffic coming to the site.

- Monetize as much of that traffic as possible by increasing the conversion rate for the site (the percentage of traffic leaving the site to a merchant site which meant revenue for Ciao).

The key business objectives that Ciao had were the following:

- Increase the size and quality of the catalog of products.

- Increase the number of offers on the web site provided by merchants and match those offers to items in our catalog.

- Increase the number of high-quality reviews written by users on the web site.

These five business objectives were mapped directly to five scrum teams. Team "Traffic" constantly monitored the amount of traffic coming to the site and did development work to try to increase traffic to the site. Team "Conversion Rate" constantly monitored the conversion rate for the site and did work to increase that rate. The "Products" team improved the quality and size of the catalog. The "Offers" team improved the volume and match rate of offers to products on the site. And the "Reviews" team found new ways to incentivize and improve the quality of reviews and the engagement of our community of review writers on the site.

Sometimes the goals and metrics one team would be going after would fundamentally oppose the goals another team would be going after. For example, the products team sometimes shipped features that increased the time users spent on the site but decreased the number of clicks that left the site to go to our merchant sites. The products team was achieving their goal of increasing the quality and richness of the catalog, so people were engaging more with that content, but then they were not moving as quickly or sometimes not at all to merchant sites to buy things which impacted the Conversion Rate team. When work one team shipped had impact on another teams' goals and metrics, the business leaders would be consulted to decide which priority won.

Keep several things in mind when creating scrum teams. First of all, try to find ways to tie scrum teams as closely as possible to business objectives and easily and regularly measured business metrics as possible. If a direct daily business metric like "conversion rate" is not calculable for a particular scrum team's metric-driven goal, try to figure out a way to quantify and score what the team is driving at. It is extremely effective to have a scrum team have a set of metrics they can constantly monitor, examine whether their work moves the metric, and if it doesn't reevaluate their work until it does move the metric. The team should set sprint-by-sprint goals for how far they think they can move the business metrics they are driving each sprint.

Table 4-1 shows some examples of concrete business metrics we have used with some of the teams we have discussed this far in the book. You can see both calculated and judged metrics in this list. Calculated metrics could be computed quite easily from data in the system. Judged metrics were more expensive as they required human judges to look at data to calculate them – which also made them so they couldn't be computed as often as calculated metrics.

Table 4-1. *Concrete business metrics*

Team	Metrics	Type of Metric
Local Data Conflation Team	Overmatch rate	Judged (weekly)
	Undermatch rate	Judged (weekly)
Local Data Web Extraction Team	Percentage of top 1000 chain businesses extracted from the Web	Calculated from catalog
	Percentage of traffic on site that shows businesses with data extracted from the Web	
	Accuracy of phone numbers extracted from the WebPercentage of businesses extracted from the Web with a phone number	
Local Data Pipeline Team	Seconds from new data appearing in a feed to being visible on the web site	Calculated from catalog and site
Ciao Traffic Team	Number of unique users per month	Calculated from site
Ciao Products Team	Number of products with pictures	Calculated from catalog
	Number of products with reviews	
	Accuracy of product-picture matching	
	Accuracy of product-review matching	
Ciao Offers Team	Number of clicks out to merchant sites	Calculated from site
	Number of products with associated merchant prices	
Ciao Community and Reviews Team	Number of community members who have posted a review deemed helpful by other community members in the last month	Calculated from site

As metrics are decided on by each team, find ways to automate the metrics. Track them on a daily basis if possible and maintain a dashboard where team members can observe whether the metrics are moving in the right direction or not.

The Power of Naming a Team   Another very powerful tool to use with teams is how you name the team. A team Eric was working with had two major problems with their product. First, it didn't *scale* properly to support many users and large quantities of data. Second, developers building new features for the product were slogging through the existing code and API complexities thereby slowing velocity in shipping new features. The teams initially had names reflecting the subsystem they owned in the product. For example, the "Graphics" team owned the graphics engine for the product. The "Layout" team owned the layout engine for the product. The team called "Graphics" unsurprisingly spent a lot of their work investing in tweaking and continuing to improve the Graphics engine, and the Layout team did the same with their layout engine. By renaming the teams to the "Scale" team and the "Velocity" team, the work output of the team naturally changed to be aligned with the most important problems the team had overall rather than the architecture blocks in the architecture diagram for the team.

When possible, it is great to have a single unifying metric that the entire team ultimately understands it is driving toward. For web properties, we have seen this number often be the monthly active user number – is this number consistently going up? On a mobile apps team, the App Store rating and the monthly active users were highly valued as the most important numbers to drive toward. For Bing's local data team, the number we thought the most about was "Q" – the composite quality score for the catalog of local businesses. But this number was somewhat flawed as it assumed that solely a highly accurate catalog would drive usage of the product – over time we discovered this metric didn't hold up very well because it was too many degrees away from the real user experience on Bing. Even with a high-quality catalog in place, there were still many issues about how and when that catalog was presented to users that made Q an imperfect reflection of the user experience with local business data on the web site.

Working with the Business to Understand Data

One of the most productive experiences we had when working on Bing's local search was to visit some of the data providers that were giving us data feeds that we combined together to create the catalog of businesses. By talking with the business people that gathered data for us, it gave important insights that allowed us to understand the data feeds that we relied on and build better machine-learned models to process and prioritize the data.

Early on in our time in Bing Local, we went on a trip to visit the headquarters of one of the major data feeds that was used to create our catalog. We had explored their data and felt that it was causing a lot of quality issues – in particular, it was creating a lot of duplicate businesses. We talked to the engineering team about their deduplication system, and it quickly became clear that the system they used was very primitive. Also, one engineering leader pointed out that because the primary customer of their data were companies that did mailings through the US postal system to other businesses, it often was "a good thing to have duplicates" because sending a couple of extra mailings to a popular business (the more popular businesses in their catalog were more likely to have duplicates) couldn't be a bad thing as it would increase the chance that your marketing mail got opened by a popular business.

Another data provider was providing us with a lot of businesses that were closed. By talking more closely with that provider and understanding how they sourced their records, we learned that a major source of their data was the scanning of phone books using OCR. In some markets, their scanning was less up to date than in other markets. In their backend system, they maintained the date that each record was scanned, but they didn't provide that data in the feed. By talking to the business people and engineers, we quickly discovered that it would be relatively simple for them to provide us with the scan date for each record in their system that was sourced by OCR. This provided a valuable additional signal that our machine learning models could use to determine the likelihood that a particular business was closed.

Data Providers   Talk with third parties providing your data and make sure you understand the way they gather data and their business model. Data providers are often willing to provide you with extra signals that they may have in their backend systems but do not provide you in their feed.

Helping the Business to Understand the Limitations of Machine Learning

It is common for a business starting out with machine learning-based solutions to expect miracles from the development team. Although machine learning can produce amazing results, it is helpful to make sure the business understands the limitations of machine learning. One of the toughest challenges we have faced with business people is helping them to understand the functional characteristics of a particular machine learning model.

Although there are many measures that can be used to evaluate a particular machine learning model such as an F1 score, we have typically used precision and recall as the metrics to communicate to the business about machine learning model performance. Here is a quick refresher that should demystify precision and recall.

Imagine a scenario where a dog catcher is tasked with going out and catching all the dogs in the city. The dog catcher fails to do this perfectly and accidentally captures some cats in addition to not capturing all the dogs in the city. Figure 4-2 shows how this story maps to the machine learning concepts of precision and recall. If each dog or cat icon represents 1000 dogs or cats, the dog catcher caught 2000 dogs and 3000 cats – but there are actually 7000 dogs in the city. The dog catcher caught 3000 false positives (cats) and 2000 true positives (dogs). So the precision of the catch is the number of dogs caught divided by the number of cats incorrectly caught plus the number of dogs caught or 2000/3000+2000 which is 40%. The recall is the number of dogs caught divided by the number of dogs caught plus the number of remaining dogs not caught or 2000/2000+5000 which is about 29%.

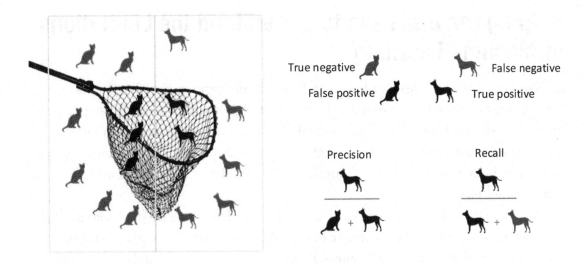

Figure 4-2. *Precision and recall for a dog catcher*

As the business begins to understand precision and recall of models which are virtual dog catchers of sorts and gets familiar with how to discuss this in terms of the precision and recall terms, you can begin to have the nuanced discussions with them about how a particular model performs. For example, business people will learn over time that it is usually impossible to have models with precision or recall approaching 100%. A model with 95% precision and 95% recall often requires an enormous amount of effort to achieve. Even a model with 99% precision sounds great, but it means that 1 out of every 100 times you are going to be getting the answer wrong. Also, if the recall of that model is 1%, it can be useless. This gap between reality and perfection is a continual challenge on machine learning teams – the business has to have a plan to deal with the times the algorithm is going to get things wrong.

In the world of local data, we dealt with the reality of models that were achieving high precision and high recall but not 100% in either in several ways. First of all, we built a system that made it very easy to quickly correct by hand any errors that were found on the web site so that the machine-learned models could be overridden when they got things wrong. We also put humans in the loop especially on data that was most likely to be shown to users. We found that although our catalog of local businesses had millions of businesses in it for the US market, 20% of what people would actually search for on our site was a set of 50, 000 entities – the most popular local businesses on the site. For those 50, 000 entities, we used machine-learned algorithms to create them, but we also used humans to check and double-check them to ensure they were 100% correct rather than 98% correct as this had a strong impact on the overall user experience since it impacted 20% of our traffic.

Another thing we found was that it was useful to talk frequently with the business about things like "the 'Closed Business classifier precision is now 97% and the recall is 85%'" because over time the business would know how reasonable it was to expect the precision and recall to improve over a given number of months of development. The business would learn that the type of work required to move precision was often different than the type of work required to move recall. They would learn about the ability to trade off recall and precision – for example, it is fairly achievable to increase recall at the expense of precision. They would learn that as models began to have precision in the 90s and recall in the 90s, that future progress would be more expensive.

The Benefits of a Strong Competitor   One advantage we had in the Bing local space was that in addition to reporting the precision and recall for our own models, we could evaluate the precision and recall for competitors. When goal setting over a 6-month period of time, we often employed a "halve the gap" rule which proved a useful rule of thumb in goal setting. In an area where we were behind a competitor – for example, we might have measured that Google's precision on phone numbers was 98% and Bing's precision on phone numbers was 94% – we found it was reasonable to plan to be able to halve the gap with Google in a 6-month period, meaning we found it achievable to set a goal to get from 94% to 96% if we had as an "existence proof" the measurement that another competitor had achieved a higher precision than us. In a vacuum, this is harder to do as you don't really know how possible it is to hit 98% precision on a particular data attribute if there is no prior art to show that it is possible to hit that high level of precision.

Communicating the Rhythm of Engineering to the Business: How We Do Scrum

Just as it is important for the engineering team to understand what is happening in the business, it is important for the business to understand what is happening in the engineering team. Failure to do this can result in the business losing confidence in the engineering team's ability to solve critical business problems in a timely way. We use Scrum and Scrum artifacts to communicate effectively with the business. In this section, we will describe in more detail the specializations we make to Scrum. This section assumes a basic familiarity with the general principles of Scrum. If you are not familiar with Scrum, we recommend The Scrum Guide by Ken Schwaber and Jeff Sutherland found at `www.scrum.org/scrum-guides/` and *Scrum and XP from the Trenches* by Henrik Knibert found at `www.infoq.com/minibooks/scrum-xp-from-the-trenches`.

The Scrum Team

For our teams at Microsoft, the product owner is typically a program manager or an architect. The scrum master is the development manager or lead or someone on the development team that is passionate about driving scrum and helping the team succeed. Whoever the scrum master is, they should expect that they can't be scheduled to do coding work at full utilization – they will need to spend some additional time each day helping unblock people and keeping the backlog groomed.

The Portfolio and Product Backlogs

Azure DevOps provides support for the scrum concepts of epics and features. These are primarily business concepts to help the business team or product owner manage the product plan. Epics and features are sometimes referred to as the "portfolio backlog." Features should be of sufficient size that they support having many user stories associated with them – a user story being the primary concept used by the development team in what is called the "product backlog." These relationships are shown in Figure 4-3.

In general, Epics transcend multiple releases and can be pursued by a team for multiple quarters (1/4 of a year segments). Features are things that typically can't be delivered in a single sprint but that can be delivered in a single product release cycle. Multiple user stories could be completed in a sprint by a five- to seven-person team, but sometimes a user story can span multiple sprints. The connection between a feature and user story is how the gap is bridged between the business-driven "portfolio backlog" and the dev-driven "product backlog"

Scrum teams will usually only talk about features during sprint planning or sprint review meetings when they are trying to figure out whether a feature is complete or not and what additional user stories have to be planned to complete a given feature.

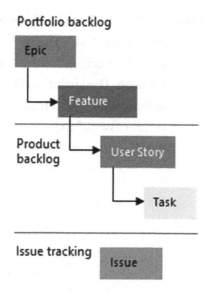

Portfolio backlog

Product backlog

Issue tracking

Figure 4-3. *Relationship between portfolio backlog and product backlog. Source:* `https://docs.microsoft.com/en-us/azure/devops/boards/work-items/ guidance/agile-process-workflow`

Table 4-2 shows some specific examples of Epics, Features, User Stories, and Tasks with rules of thumb about typical durations and typical number of these per parent type for a hypothetical team that is working on allowing users to search Office 365 content (documents, contacts, emails, calendar) within Bing.

Table 4-2. *Epics, Features, User Stories, and Tasks*

Concept	Example	Typical Duration	Typical Number of These per Parent
Epic	Allow users to search their Office 365 data in Bing	Multiple releases (quarters)	Five to ten total epics for major releases
Feature	Allow users to search their Office 365 contacts in Bing	Release (months)	Fifties to hundreds of features in an Epic
User Story	Build experience to allow users to browse their Office 365 contacts	Sprints (one to three sprints)	Tens of user stories to finish a feature
Task	Write code to get contact XML from Office 365 graph	Days (one to five-ish days)	Tens of tasks per user story

Our product backlog is composed of Azure DevOps "User Stories" as shown in Figure 4-4 which are in turn broken down into tasks. The product backlog records everything the team wants to do. The sprint backlog (under "sprints" then "Backlog" in Azure DevOps) shows the sprint backlog. The sprint backlog is created as part of the sprint planning meeting and can ideally just be pulled from looking at the top-priority user stories in the product backlog (under "Work" then "Backlog" in Azure DevOps) and adding them to the next sprint in priority order until the sprint is full.

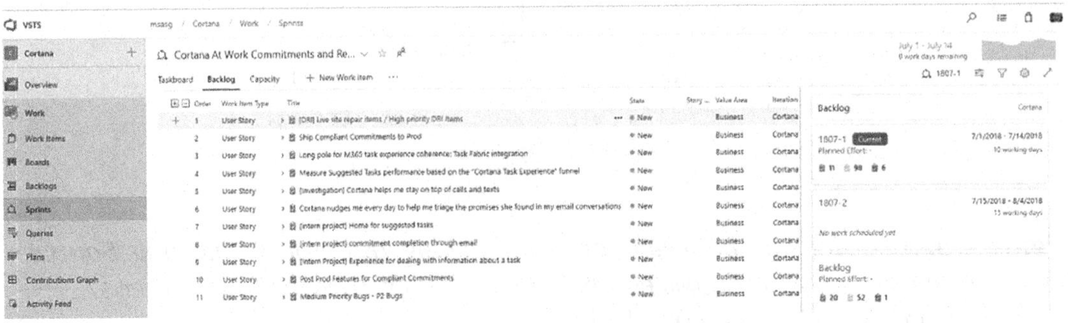

Figure 4-4. *Example Azure DevOps user story backlog*

User stories in the product backlog shouldn't be massive. As a rule of thumb, they should be completable within one but at most within three 2-week sprints. As an estimation of whether your granularity for user stories is right, a typical scrum team of five to six people should be able to take on roughly seven user stories during a sprint. This could of course vary, but shoot for having user stories that aren't too big and aren't too small.

Jira and Other Models of Driving Scrum   Although this section describes Azure DevOps as an environment to drive scrum within, all these principles work within Jira as well which Eric used extensively while at Ciao. It is also worth mentioning that much of this can work in a paper-based system that is done on physical boards. The aforementioned *Scrum and XP from the Trenches* book describes a paper-based system. We have done this as well although depending on where the business people are located, you may have to take daily photos of the physical boards or transcribe key information and send it out in email regularly to apprise the business of your progress.

User stories can live on from sprint to sprint – but they should be completable within a small number of sprints if possible, ideally one. If user stories are too big to complete in a small number of sprints, this is a good sign that they should be further decomposed. Look for user story names that have "and" in them – this is a good sign that the user story should be split into two items.

Backlog Scrubbing   A best practice for managing the product backlog is to "scrub" the backlog frequently. Scrubbing involves continuously looking at the top "tens" of user stories on the product backlog and making sure they are prioritized correctly and well described. Some teams schedule weekly or biweekly meetings to scrub the backlog – this can help sprint planning meetings go faster.

User Stories

Figure 4-5 shows an example User Story in Azure DevOps.

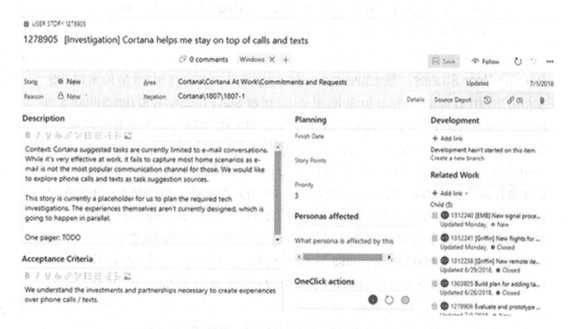

Figure 4-5. *An example User Story in Azure DevOps*

User stories can be optionally associated with a Feature – this ties the "product backlog" to the "portfolio backlog" and helps the PM team track progress in their part of the world which is the portfolio backlog. User stories can be created by anyone, but only the product owner sets the priority, and only the development team estimates the size and adds a user story to a sprint.

A user story has many fields including the following:

> Feature Backlink: This indicates for which feature in the portfolio backlog this story is being implemented. This isn't required if a user story has no connection to a feature – for example, the scrum team might decide to do some work on its own developer tools or DRI tooling, and this would not have to be parented to a feature if the PM team doesn't really care about it.

> ID: A unique identification, just an auto-incremented number. This is to avoid losing track of stories when we rename them.

> Name: A short, descriptive name of the story. The name should be specific enough that developers and the product owner understand approximately what we are talking about and clear enough to distinguish it from other stories.

Naming User Stories   Because we encourage the business team to look at the scrum board, it is important to at least write user story names and descriptions in a way that anyone in the business can understand them by avoiding jargon and describing the business result we are trying to achieve. So instead of having a user story that says "Generate better n-grams to change Match Function to Improve Match Rate," we would write it as "Reduce the number of duplicate businesses with better word separation." Tasks under the user story can and should be technical and use jargon as they are for the scrum team in tracking itself – the only important information the business team gleans from tasks is who is working on a user story, how much work remains, and how much work has been completed.

> Description: Can in many cases just be the name again or a more detailed explanation of the user story. Once again, this needs to be written in language that people outside the team can understand it.

Business Value: The product owner's importance rating for this story. In Azure DevOps, you indicate this by moving user stories up and down vertically in the backlog by dragging them.

Story Points: The unit is story points, and for our teams we have this correspond roughly to "typical man-days." Ask the team, "If you can take the optimal number of people for this story (not too few and not too many, typically two) and they have a typical day with interruptions and meetings where they probably have about 5 hours in the flow coding, after how many days will you come out with a finished, demonstrable, tested, releasable implementation?" If the answer is "With three people it will take approximately 4 days," then the initial estimate is 12 story points.

Once you have an initial story point estimate to a user story, you don't really worry about it again. This is your initial "best guess" and is primarily used during sprint planning to estimate how many user stories you can load into a given sprint. Once a sprint is underway, tracking of the cost of a user story moves from story points to breaking the user story down into tasks that are costed in ideal dev hours (we assume on our teams that a developer can do 5 hours of work per day when including interruptions, meetings, distractions, etc.)

Story Points   For scrum purists, our adulteration of what story points are is probably offensive. The way story points are supposed to work is the team learns over time how many story points they can achieve in an iteration and the unit is flexible and the team gets a good sense of how much work a ten-story point work item is. But we have found this hard to use in practice with large teams having several different scrum teams within them – it is also hard to explain to the business that "Team A's story point size is different than Team B's story point size." So we have not been able to have story points be flexible units in practice and have instead fallen back to the "a story point is a typical man-day of work" and "when you get to task breakdowns, you can do 5 hours of work in a day."

Acceptance Criteria or "How to Demo (HTD)": This is a high-level description of how this story will be demonstrated at the sprint demo meeting. This is essentially a simple test spec. "Do this, then do that, then this should happen." With TDD (test-driven development), this description can be used as pseudo-code for your acceptance test code. This is often put in the "Acceptance Criteria" field but can also be in the description – just mark it with "HTD".

User stories can be assigned to someone – typically the person most knowledgeable about the story who could answer questions about it from outside people. But in practice, a lot of user stories never get assigned to anyone. This is fine as the real "work" on a user story is tracked by tasks, and tasks are always assigned to individual developers. So if the user story is not assigned, you can see in the task board all the developers that have worked on it so far as there will typically be multiple tasks with a variety of devs on them.

Tasks

Figure 4-6 shows a Task associated with a User Story. Tasks can be very lightweight and usually just capture these elements: Title, Assigned To, Time Remaining, Status (New/Active/Done), and "Which User Story is this associated with." Tasks are used to coordinate who is working on what in a User Story and break down a User Story into the logical tasks it takes to complete that story. We don't use the "Estimate" field. Description could be used to capture additional detail – but this is only if it is useful to do so for the developer working on the tasks. Mostly a developer uses a task as just a simple "TODO" for work they have to code up, and mostly they understand the work from the task description. Here we don't care about whether the business people can read the task or not – the customer of the task is the developer working on the task.

We use the Time Remaining field in this way: when a task is first assigned to a dev to work on (this is all done in the daily standup or sprint planning meeting), the dev does a quick estimation of how much work it is in hours – for example, the dev might say, "This will probably take 10 hours" (e.g., 2 dev days). This initial estimation should go in the "Time Remaining" field, not the estimate field. Then with each subsequent daily standup, each developer with an active task reestimates how many hours are remaining for each active task – for example, "This task still is looking like I have 10 hours remaining; it was bigger than I thought" or "This task only has 3 hours remaining; it was easier than I thought," or "This task is done" in which case the task is just moved to the "Complete" column.

The example task in the following illustrates that often the first task in a user story will be to assign an expert in the area to "create a plan for how to deliver this user story."

Figure 4-6. *A simple task*

Task Linking to Pull Request in Git

At Microsoft, we use Git as our version control system. It is valuable to track the pull requests that result in tasks being completed. When a developer makes a pull request, he or she can specify in that pull request all tasks that are completed once that pull request is merged. This allows auditors in the future to connect our backlog to the development work.

Bugs

Any user story can have a bug logged against it as shown in Figure 4-7. When a bug is created against a user story, that bug almost always has priority over other tasks in that user story. The principle is to fix the existing bugs before advancing the state of the user story.

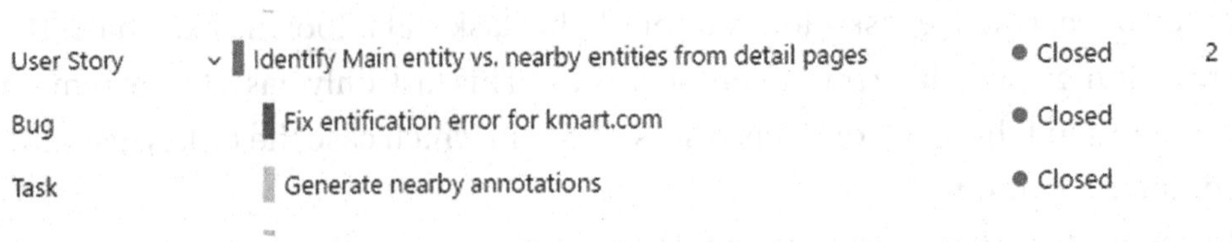

User Story	⌄ ▌ Identify Main entity vs. nearby entities from detail pages		● Closed	2
Bug	▌ Fix entification error for kmart.com		● Closed	
Task	▏ Generate nearby annotations		● Closed	

Figure 4-7. *User Stories can be associated with both tasks and bugs*

If there are bugs filed against already shipped user stories that are P0 (Priority 0) bugs, these should be brought into the sprint immediately – they are considered sprint breakers. When possible, the already shipped user story should have a bug added to it and then be moved into the current iteration path for the sprint – but mostly this doesn't happen in practice. We will describe the approach that seems to work more in the following. When the scrum team has their daily standup, they should look for these new high-priority bugs and immediately schedule the work into the sprint to fix them.

Lower-priority bugs should be considered during the next sprint planning meeting and brought into the sprint if they are deemed critical to fix in the subsequent sprint.

It is hard once a project is moving to attach bugs to user stories that have already been shipped. So the usual approach we see teams use is they have a "High-Priority Bugs," "Medium-Priority Bugs," and "Low-Priority Bug" User story that they create in a sprint as shown in Figure 4-8. They then associate incoming bugs with one of these three user stories and prioritize the work appropriately – for example, they first try to complete high-priority bugs, then high-priority user stories, then medium-priority bugs, then medium-priority user stories, and so forth.

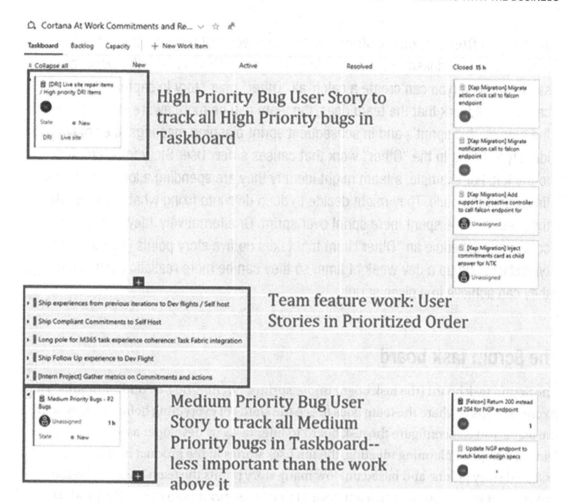

Figure 4-8. Tracking Bugs alongside work in an Azure DevOps task board

Tasks That Don't Fit into a Story   Sometimes you will have work that the team is doing that just doesn't fit in an existing user story. This is fine – the work should still be tracked (you can create a catch-all "Other" user story to capture it). By capturing all work that the team does, the team can be more aware of "where did the time go this sprint"; and in subsequent sprint planning meetings, they may identify a theme to the "Other" work that causes a new User Story to be created to track it. For example, a team might identify they are spending a lot of additional time fixing the build. They might decide to deep dive into fixing what is wrong after they see the time spent there sprint over sprint. Or alternatively, they might just continually schedule an "Other" item that takes up five story points (if Other was typically taking up a dev week of time) so they can be more realistic about what they can achieve in a given sprint.

The Scrum task board

The Scrum task board (the task board under sprints, not the "Board" under "Boards") in Azure DevOps is where the team goes to see the status of everything being worked on by the team. You can configure the task board to filter tasks by developer assigned to them. During the sprint planning meeting, the top user stories in the product backlog are assigned story points; and based on how many story points the team has learned they can commit to for a sprint, the team brings the highest-priority stories in the product backlog into the new sprint backlog. Once the stories are in the sprint, the team refines them further by breaking each story into tasks (and the initial story point estimates are ignored going forward but can be observed later to see how good the team is getting at initial estimates).

Tasks are easily added to a user story in the board by clicking the "+" button to the right of a user story in the task board. Tasks begin to flow when they are dragged horizontally from the "New" column to the "Active" column. Tasks are always assigned a developer when they are moved to the Active column and always assigned a "Time Remaining" in hours (where 5 hours = 1 dev day). Tasks in the "New" column can be preassigned to the dev that will do them and pre-costed, but this is not necessary and is timeboxed by the sprint planning meeting and time available in sprint standups.

As tasks are completed, they are dragged horizontally to "Closed" when the task is verified "Done." We don't use the Resolved column for User Stories; it is for bugs. The Scrum task board is shown in Figure 4-9.

Note that the "Resolved" column is used instead of the "Closed" column when a Bug fix is completed in the sprint backlog. The Bug is moved to Active when a developer starts working on it. When it is fixed, tested, checked in, and deployed to production, the bug is moved to "Resolved". The original person who opened the bug should verify that it was fixed and then move it to Closed.

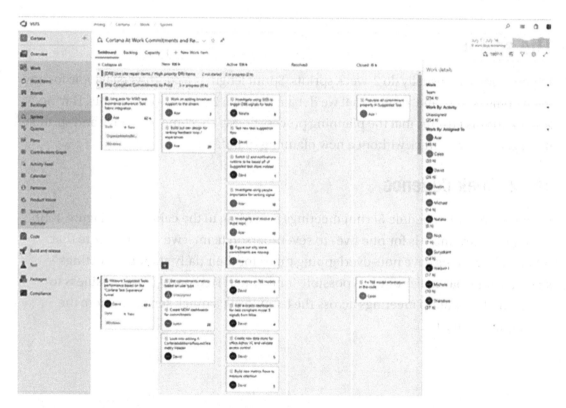

Figure 4-9. *The Scrum task board*

We use Azure DevOps's task board to be the single status page for each team. This page includes a Burndown graph (in the top right corner) that is automatically kept up to date when teams move and update their tasks regularly: updating Time Remaining for a task when it is first moved to the Active column and updating Time Remaining for active tasks every day thereafter until it is done and moved to the Closed column.

All of this is timeboxed (an Agile term meaning completed in a fixed period of time) to the standup meeting – no task updates need be done outside the standup. The scrum master is the servant for the team in helping the team keep the task board up to date so it doesn't impact them outside of the standup. During the daily sprint meeting, the scrum master projects the scrum dashboard; and each participant can report the Time Remaining on their work items in progress, and the scrum master updates all active tasks while projecting (update Time Remaining in the bottom right corner, assign new tasks by dragging vertically, mark tasks as done by dragging horizontally).

The sprint

For our teams, we usually do 2-week sprints. Sprint length is a controversial topic for any team doing scrum, and we find that we debate between 2- and 3-week sprints. Having 2-week sprints ensures that the planning process doesn't become too heavy and sprints don't get too far off track without a new planning round.

The 2-Week Cadence

On our teams, we schedule Scrum meetings according to the calendar in Figure 4-10. This specific example is for one five- to seven-person team – we like to ensure that other sibling teams have non-overlapping times for their daily standup meetings so that on any particular day it is possible for an interested party in the business to attend all the standup meetings across the team to get a quick sense of where the development effort is.

M	T	W	Th	F
First day of Sprint 10:00-11:00— Sprint Planning Meeting	10:45-12:00— Daily Standup	10:45-12:00— Daily Standup	10:45-12:00— Daily Standup	10:45-12:00— Daily Standup
10:45-12:00— Daily Standup	10:45-12:00— Daily Standup	10:45-12:00— Daily Standup	10:45-12:00— Daily Standup	Last day of Sprint 11:00-12:00— Sprint Review and Retrospective 12:00-1:30— All-Hands Sprint Demo Meeting

Figure 4-10. *The Scrum Calendar*

Daily Scrum Standups

Daily Scrum "standups" will be scheduled in a way that two standup meetings aren't running at the same time according to this schedule. Standups are timeboxed to no more than 15 minutes.

The daily standup is done in a room with a projector or TV or Surface Hub, and the scrum master projects the Azure DevOps task board. The task board is shown in turn filtered by each developer in the meeting. Each developer in the room can then refer to their active tasks as they answer the Scrum questions "What did I do yesterday?" "What will I do today?" and "Am I blocked?" The developer then says for each task either "that one is done" in which case the scrum master immediately drags the task to the Closed column, or the developer estimates the number of hours remaining for active tasks (if it has changed since the last standup), and the scrum master immediately updates the Time Remaining for the task (in the bottom right corner of the task).

If a developer has completed all active tasks, tasks previously assigned to the developer but not yet started are dragged from the "New" column to the "Active" column, and the developer estimates the number of hours remaining. The scrum master puts the number of hours remaining in the bottom right corner of the task.

Alternatively, a developer may have no unstarted (aka New) tasks assigned, so tasks are grabbed from the "New" row or from other developers that need help on their tasks. Any time a task is moved into the Active column, the initial Time Remaining must be put in the task.

Updating the task board Outside of Standup   We actually ask our developers to not update the task board outside of the daily scrum standup meeting because the updating of the task board (this task is finished, this task has less hours remaining on it now, I am taking this new task) is fundamental communication to the rest of the team, and if this communication happens out of band from the standup meeting, the communication of each developer's progress is lost to the team.

The standup should also be used to triage blocking issues the team has. The standup should not spend too much time discussing these issues unless they can be quickly resolved – remember the 15-minute timebox. If an issue requires more discussion, the scrum master sets up another time later in the day to discuss it with the involved parties

End of Iteration Meetings

We like to have all our sibling teams start and stop their sprints on the same day, and we combine some meetings across our various five- to seven-person sibling teams to have larger demo meetings. So for a typical team of around 50 developers all working on the same project – for example, Bing's local data team – we would have around six scrum teams that would participate together in a combined all-hands sprint demo.

All-Hands sprint demo

On the last Friday of the sprint, the All-Hands sprint demo is held as one combined 1.5-hour meeting. We do this meeting in person, but we also record the meeting so that it acts as another artifact of scrum. The format of the All-Hands sprint demo is as follows.

Each team is timeboxed according to the number of devs on their team – each team gets 2 minutes per dev. So if one team has seven devs, they get 14 minutes to demo, another with five gets 10 minutes, and so on. Demo times are strictly timeboxed by Eric's dreaded countdown timer clock shown in Figure 4-11. When the clock beeps, whoever is talking completes their sentence and hands the meeting over to the next team. We found a timer was necessary to ensure that the demo meeting didn't drag on – having the timer makes sure that everyone is prompt and presents their work quickly.

However, if someone asks a question or has a question or comment during the demo, we give time back to the countdown as we don't want to discourage good comments, questions, or suggestions. The scrum master has power to give back time on the countdown and also suggests that out-of-control discussions be taken offline.

Figure 4-11. *A countdown timer clock – used to timebox sprint demo meetings*

Some words on logistics to make this work – we have followed this process for teams with up to 50 developers (the meeting lasted about 1.5 hours and had six different teams presenting). To ensure that the meeting goes smoothly, we do the following.

An all-hands sprint demo deck is prepared in advance of the meeting that is shared by all the scrum teams. Each scrum team has a "cover slide" that shows their sprint goals, whether or not they met them, and key metrics for the team, and what improvements they made to those metrics during the sprint. The team can then add their own slides after their cover slide for specific work they are going to demo.

---

sprint goals   We are very transparent about what the sprint goals are and whether or not they are being met. As you will see later in this chapter, the sprint goals and whether they are met or not are broadcast to the business team and other stakeholders. Communicating sprint after sprint that you aren't hitting sprint goals can be a good signal to other teams that they can't trust that you will complete the work you say you are going to complete – so you need to continually emphasize that teams need to not think of sprint goals as being too aspirational but achievable. Eric has started calling sprint goals "Sprint Commitments" to further press home the notion to his team that these need to be thought out and costed to ensure they can be met.

Where possible, demos are prerecorded and can just be linked to from the slide deck. However, we encourage teams to not spend a ton of time preparing for the demo meeting, so ad hoc demos are also frequent.

We use conferencing software like Skype or Microsoft Teams to project the demo deck; and users presenting ad hoc, video, or slide-based demos can connect to the conference in progress and share their own screen while demoing.

Sprint Retrospective

The sprint retrospective meeting is usually scheduled for the last Friday of the sprint and is not a combined meeting – this is just done separately for each of the smaller five- to seven-person team. This meeting acts as the final "standup" for the sprint as well in which tasks are closed out, user stories that are not yet completed are moved to the next sprint, and the backlog is reviewed and reprioritized.

We try to add at least one user story to the new sprint that is a direct result of a suggestion for an improvement in the retrospective section of the meeting.

We also like to roll up to all the other teams in email a report of everyone else's retrospective meeting so that the rest of the team can at least read about how the other teams felt about their sprint.

Sprint Planning Meeting

The sprint planning meeting is scheduled for the first Monday of the sprint according to this schedule. This meeting is staggered where possible so external "chickens" can observe – but remember the scrum principle of chicken and pigs.

Scrum Chickens and Pigs   The chicken and pig story relates to a chicken and a pig that want to start a ham and eggs breakfast restaurant. The pig begs out because he'd be **committed** to the restaurant, but the chicken would only be **involved**.

Rather than having the sprint demo meeting be the meeting where product backlog suggestions are made, we use the sprint planning meeting, although discussions can happen during the sprint demo meeting that lead to ideas for the backlog.

The sprint planning meeting is typically conducted in a conference room with a projector and is timeboxed to 1.5 hours. All the development team are present. Non-development team members are also welcome to observe, but their participation should be limited. If there isn't enough seating, pigs sit at the table, and chickens sit against the walls.

Ideally, the product backlog has already been scrubbed prior to this meeting – this will make this meeting go faster if the product backlog is already prioritized, defined, and ready to go.

First in the sprint planning meeting, the previous sprint is closed out, and any outstanding user stories are moved from the old sprint into the newly created sprint.

If the product backlog isn't already scrubbed, the team spends some time looking at the product backlog and ensuring that all high-importance user stories that have been thought of are in the backlog and prioritized correctly. This is a point where "chickens" can chime in and suggest user stories that potentially the team hasn't thought of but need to do. The product owner is very influential in this discussion and helps the team understand what needs to be accomplished next.

Sometime during this meeting, the team should also take some time to try to determine what their sprint goal or goals are for the sprint. See the section "sprint goal" in this document on why a sprint goal is important.

The team then fills out the "Capacity" tab in Azure DevOps's "sprint" section. At this point, the team determines vacations, whether everyone in the scrum team is full time on the sprint this scrum, or whether some people are only partially on the sprint and partially on another scrum team (which is undesirable but sometimes happens). Team members can also log their vacation for the sprint. Figure 4-12 shows a filled-out capacity page. The maximum capacity per day for any developer is always capped at five. Looking at the Capacity tab for this sprint, you can tell at a glance that

Azar is taking a week off during the sprint.

Michael and Michael aren't going to participate in this sprint – their capacity is set at 0 (maybe they are in the list because they have participated in past sprints).

Nick will only be able to participate at about half-time during the sprint (in this case because he is a lead).

The team has 1 day off – maybe a cool morale event!

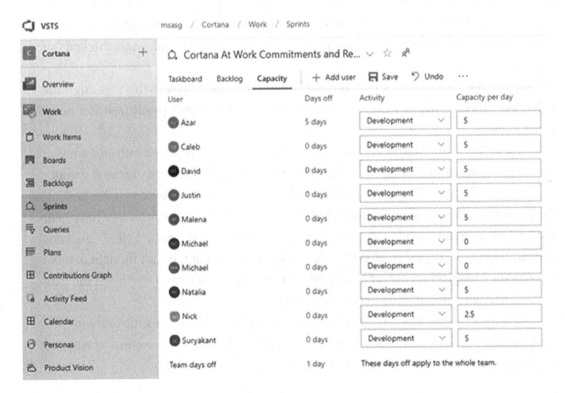

Figure 4-12. *Azure DevOps Capacity tab*

Once capacity is set, the team immediately knows how many story points they can take on in the sprint (total capacity in hours divided by five). The team then looks at the top user story and estimates its size in story points (one story point equals 1 dev day or 5 hours).

Planning Poker   One useful technique of estimating the size of a user story is a technique called planning poker. The main idea of planning poker is to let multiple team members simultaneously estimate the cost of a user story and then find who estimates high and who estimates low on the team and determine what assumptions went into the high and low estimates. Often someone with a high estimate will have thought of something the rest of the team hasn't thought of that needs to be considered. Also, someone with a low estimate may have an idea for doing the work that is much cleaner and simpler than what everyone else thinks has to be done. For more on planning poker, see this blog post: `www.mountaingoatsoftware.com/agile/planning-poker`.

Once the capacity of the team is determined and the existing stories that didn't get completed in the previous sprint are deducted from the capacity for the new sprint, additional stories to fill the sprint's capacity are added in the priority order that they are listed in the product backlog. The team doesn't need to estimate story points for every user story in the product backlog, just the ones that will fit into the sprint backlog plus maybe a few more for buffer in case some stories complete earlier than expected and new stories are added from the product backlog to the sprint backlog later in the sprint. All User stories in the sprint backlog must have an initial story point estimate.

With the sprint backlog filled with enough story points to cover the capacity of the team, the team then works to break down each user story into more granular tasks if there is remaining time. They do this by adding child Azure DevOps tasks to the user story. Some user stories may already be small enough that they are only a single task – this is OK, but just create one child task for the user story with the same name as the user story. When the timebox for the sprint planning meeting is hit, any remaining user stories that haven't been broken down yet can be assigned to individual team members to break into tasks later on during their own time.

Communication of Scrum Status to the Business via Email

We send two regular emails to the business, one after the Friday end-of-sprint All-Hands sprint demo meeting that summarizes what has been done for the past sprint and one after the Monday sprint planning meeting for the new sprint which summarizes what will be done in the new sprint.

End-of-Sprint Email

The end-of-sprint demo meeting email contains the following:

- A link to the video of the All-Hands sprint demo meeting

- A link to the slide deck used in the All-Hands sprint demo meeting

- A brief "CliffNotes" version of the meeting that includes in written form which goals were met and not met and what was demoed in the meeting along with time markers to the video where the demo happened.

An example end-of-sprint demo meeting mail looks something like this:

Hello—this is the sprint 8 summary for Local Data which completed on Oct. 25. The video of our demo meeting is here and the slide deck is here.

Here is what our various teams completed in this sprint:

Pipeline Team

Sprint Goals Completed

- Local Probe V.Next is deployed, supports switching between last 5 full runs and searching identifiers
- Publish pipeline optimization (merging jobs and moving to Common Data Access Layer) reduces runtime by 1.5 hours
- Markets can now configure unstructured address geocoding with their own market-specific logic. The ko-KR team used this to fix 88% of bad geocoded Korean addresses.

Not Completed

- Neighborhood cache refresh to enrich 2.1M more entities with neighborhood information for EN-US

Demos:

Local Probe v. Next Demo (5:15)

Local Probe v.4 has been deployed to autopilot. Old local probe is now defunct.

You can now switch between the last 5 full runs.

You can also search by identifier. For example "1x1" lets you search the identifier list for the sentinel entity.

Publish Pipeline Optimization Demo (8:30)

We integrated two more jobs into the publish job. We merged LesGeoOntoloty and Neighborhood jobs into the PublishIndex job. We also migrated the content generation in PublishIndex job to use the common DAL.

This work has reduced about 1.5 hours off the publish pipeline runtime. The new publish pipeline runtime is around 5 hours.

Conflation Team

Sprint Goals Completed

- Definitive Feeds Design Proposal
- Corrections Dashboard ships which allows multiple views of corrections including feeds that are being corrected, attributes being corrected, and a way to validate that corrections on an entity have been applied

Not Completed

- ID churn dashboard

Demos:

TCL Model Bin File Automation (11:15)

We can now automatically prepare TCL bin files during build.

Definitive Feeds Design Proposal (14:29)

We presented a proposal for a new feature called definitive feeds.

Motivation: consider a query like Starbucks in Bellevue. We display way more Starbucks than their actually are due to old/bad data. Yet we can wrap the Starbucks web site and know exactly the current set of Starbucks. What we need is a feature that can suck in the definitive feed of Starbucks from their web site and publish those while simultaneously deleting all other Starbucks that don't match with the definitive feed.

Beginning-of-Sprint Mail

The beginning-of-sprint mail contains the following:

- A link to the task board for each team

- A list of the team members on each team and their capacity for the sprint (via a link to Azure DevOps capacity page for the team for the sprint)

- An "English language" capacity for the business – telling how many dev days of capacity the team has available in the next sprint

- High-level sprint goals for the new sprint

An example beginning-of-sprint email looks like this:

Hi all. Platform team completed our planning meetings for the 1812.1 sprint (December 3-December 14). Here are the planning results and the sprint task board links and sprint goals for all the teams.

Pipeline Team

Sprint task board is linked *here*
Scrummaster: Taylor
Team Members: Abram, Joan, Dave, Emma, Patrick, Shahar, Stan (65 dev day capacity)

Sprint goals:
Get Mexico market up and running
Refresh the neighborhood cache
Reduce overall size of cache files

Conflation Team

Sprint task board is linked *here*
Scrummaster: Matt
Team Members: Ade, Max, Linus, Amy, Leah, Mike (45 dev day capacity)

Sprint goals:
Ship the ID Churn Dashboard
Ship Mexico conflation model
Move force match to run later in the pipeline

Retrospective Mail

Another mail that we have occasionally sent to the business is a report of the retrospective. In our retrospectives, we gather up verbatim "Good," "Bad," and "Meh" comments along with specific user stories that were arrived at to schedule as a result of the retrospective. Here is an example of a retrospective readout for just one of the local data teams after a sprint:

Hi all, the Pipeline team just completed their retrospective for this sprint and here are the results:

Pipeline Team Retrospective

Good
Holidays
Excited about codegen
Gathered more team goals/metrics
Deletion PRs have gone in
Good communication with Conflation team lead to two features getting added to the platform
Code reviews lead to helpful feedback and were dealt with quickly
Felt productive
Major progress on inval batching design
Wrapping up pipe event work, excited to move to tooling
Have a new customer for pipe deref
Productive meetings

Meh
Lingering build/infra issues
Lots of long-tail work remaining from previous teams

Bad
Common workflow errors go unaddressed because no one complains directly
We still need to improve build "team" workflow/responsibility delegation
Forgot to use test-driven development and it lead to wasted time
No good process for getting in large PRs w/ constant merge conflicts
Builds failing all the time for sometimes absolutely no reason
"Hang bug"

Follow-up

Get our perf tests running on the CI server

Better/more thorough bug triage at least once a week

Figure out how to prevent asserts from hanging the CI server

Publish a wiki that lists which configurations are supported/work

Come up with plans for reducing memory usage short-term

Conclusion

In Chapter 4, "Aligning with the Business", we have discussed the importance of communicating with the business team. We have described how important it is to have teams that are aligned as directly as possible to business metrics and business goals. We have shared some thoughts on how to work with the business to help them understand the limitations of machine learning. We have also described how we do Scrum and how we involve and communicate to the business team through scrum meetings, scrum artifacts, and emails around our scrum cadence.

In Chapter 5, "Motivated Individuals", we will consider the importance of building projects around motivated individuals and how to increase the number of motivated individuals within a team.

CHAPTER 5

Motivated Individuals

*Build projects around **motivated individuals**. Give them the **environment** and **support** they need, and **trust** them to get the job done.*

—agilemanifesto.org/principles

Early in the history of Bing's local search product, we had a much more primitive conflation/merge engine that was costly to maintain and produced mediocre results. Several individuals that worked with that conflation engine day in and day out felt there was a better way to do conflation. The existing conflation system had become a bug farm and required a lot of work to keep it working. The new proposed approach seemed promising, but the challenge was to find time to build something new while continuing to "keep the lights on" in the old system.

Fortunately, the entire data pipeline was designed in a componentized and modularized way with a well-defined input boundary to the conflation engine (a set of XML files that the conflation engine read) and a clearly defined output boundary leaving the conflation engine (a different set of XML files the conflation engine wrote). This made it possible to replace the conflation engine without disrupting the rest of the system.

Two motivated individuals were empowered to drop all their work in the current system and go off on their own to develop a new conflation engine. They were "ring-fenced" – a term from finance that is used at Microsoft to mean making sure that the ring-fenced team is isolated from the main team so they aren't paying any "taxes" in their development time to support the current active production project. Management was informed to expect slower progress for the 6 months that a new conflation engine was under development. Although initial estimates were more optimistic than the actual 6 months the new development took, regular reviews of the progress of the new project indicated that it was on track. Management provided needed air cover to buy the required time for the new conflation engine to be completed. As the two motivated

© Eric Carter, Matthew Hurst 2019
E. Carter and M. Hurst, *Agile Machine Learning*, https://doi.org/10.1007/978-1-4842-5107-2_5

individuals neared the finish line, additional team members were added to help do additional integration work around the new conflation engine to enable it to be used in production.

The old conflation engine and the new conflation engine ran head-to-head for several weeks which exposed bugs and issues in the new code that were quickly fixed. As the quality of the output of the new conflation engine began to exceed the old conflation engine, the switch over to the new conflation engine was completed with major gains in the quality of our conflation in the local data system.

But the critical success factor in this project were the two motivated individuals. They spent a ton of energy and effort to ensure the project would be a success. They felt empowered to go fix things that were major problems in the system. That energized and motivated their work and led to a successful project result.

Rewrite Frequently

In Bing, we strove to engineer systems to allow them to be rewritten frequently. This is a critical capacity for an agile team – it was completely doable to throw away the existing conflation system and write a new one because it was loosely coupled with the rest of the data pipeline. Also, it was easy to rewrite a new system and run it side by side with the old system – so-called A/B testing – to ensure the new system could run as successfully and with the same or better level of quality than the old system.

Fergus Henderson in his paper "Software Engineering at Google" observes that Google tries to rewrite most of its software every few years for similar reasons:

> *This may seem incredibly costly. Indeed, it does consume a large fraction of Google's resources. However, it also has some crucial benefits that are key to Google's agility and long-term success. In a period of a few years, it is typical for the requirements for a product to change significantly, as the software environment and other technology around it change, and as changes in technology or in the marketplace affect user needs, desires, and expectations. Software that is a few years old was designed around an older set of requirements and is typically not designed in a way that is optimal for current requirements. Furthermore, it has typically accumulated a lot of complexity. Rewriting code cuts away all the unnecessary accumulated complexity that was addressing requirements which are no longer so important. In addition, rewriting code is a way of transferring knowledge and a sense of ownership to newer team members. This sense of ownership is*

crucial for productivity: engineers naturally put more effort into developing features and fixing problems in code that they feel is "theirs." Frequent rewrites also encourage mobility of engineers between different projects which helps to encourage cross-pollination of ideas. Frequent rewrites also help to ensure that code is written using modern technology and methodology. `(https://arxiv.org/ftp/arxiv/papers/1702/1702.01715.pdf)`.

Rewriting frequently motivates individuals and shows you that you trust them. Individuals are not motivated working on old code that they don't understand, that they didn't write, and that they feel no connection to. Aim to rewrite frequently to create more motivated individuals in your organizations.

Finding and Generating Motivated Individuals

An organization needs to work at not just finding but generating motivated individuals. The first source for motivated individuals should be within your existing organization. We have often seen organizations that seem to systematically demotivate individuals. How can you ensure that you generate a constant flow of motivated individuals within your current organization?

Perhaps the most critical thing to do to generate motivated individuals is to create compelling goals, metrics, and targets – communicate a very clear vision of what the team needs to achieve and provide a set of metrics that will measure progress to achieving that vision – but leave the "how" to the team and even more importantly communicate your trust that the team will figure out the how. This goes to the "trust them to get the job done" part of the manifesto.

As an example of this, we worked for a while with a sister team. In that team, there were one or two people who generally "called all the shots" for the organization. Everyone else in the organization was expected to follow the game plan set by the few. By comparison, our team encouraged many people to chime in and share their best ideas and best strategies. Our team left the "how" to the many and trusted the team. The other team left the "how" to the few which didn't convey trust to the rest of the organization. Because the leaders of the autocratic team were quite brilliant, we can't say the other team wasn't successful – they pursued a lot of good ideas and strategies. But we believe the people on our team felt more trusted and empowered and certainly more motivated.

For a manager or leader in a team, a good way to think of this is what we call the "Peter Jackson Approach to Project Management." For the *Lord of the Rings* movies, Peter Jackson had a vision of what he wanted the movie to accomplish and be.

To communicate this vision to his team, he worked with a storyboard artist to provide a set of storyboards that covered the whole movie. He said:

> *[Storyboards] are a cheap pass at the movie. I get to make the movie at a really, really low cost, for the price of a few pencils and some paper, but it has effectively put me through the process of making the film. As a director, I've had a go, I've done version number one, and I can get to look at the movie complete.*

His producer Barrie Osborne added:

> *[The storyboard] was a great tool. It was never intended and never was the final version of the movie. I've worked with some directors who will storyboard their film and you'll really see those storyboards on screen. Peter is not that kind of a director. Peter uses it for inspiration, for communication with the departments, but you know that those storyboards are going to change, they are a starting point and not an endpoint for Peter.*

The storyboards created by Peter Jackson served to help the team understand very clearly "what" it was they were trying to create. But the "what" was created in a low-cost way – with simple pencil drawings. The "what" was also used to get everyone on the same page, but the storyboard was by its very nature a low-cost simple pencil drawing that was easily modified and could be easily discarded if it wouldn't work or a better idea was thought of. Creativity of individuals on the team was encouraged by the light-weight vision represented by the storyboards, changes could be made to the plan and were. But having an initial plan was essential.

To apply the Peter Jackson school of project management to data engineering projects, find ways to create "storyboard-weight" vision documents that are easily thrown away, modified, and primarily used to communicate the vision to the team. These can often be oriented around a future customer experience. You can, for example, storyboard the experience a customer will have with your product a year from now. As the team gets a clear idea of what the product is envisioned to be a year hence, they can rally around the "how" of making the product vision come to life. By not dictating the "how" to the team, motivated individuals naturally emerge who not only come up with innovative "how's" but also feel empowered to change the "what" when they see the vision is lightweight and open for innovation. Just be careful to make the storyboards a "starting point and not an endpoint."

Don't Forget the Metrics   Always combine "storyboard-weight" vision documents with metrics that measure what you are trying to achieve with your product. If there is one paramount message in this book, it could be expressed as "It's the metrics, stupid!" Metrics can also counteract the tendency the team will have to put too much "how" into the planning documents and vision. For example, Amazon's project to try to deliver packages to houses with drones probably has a metric somewhere like "minimize the time it takes from order to delivery" and another metric somewhere like "minimize the cost it takes to deliver an order." Should someone on the team come up with an alternate "how" that beats drone delivery for those metrics – in-home 3D printing? Teleportation? – the metrics trump the vision of drone delivery, and the new "how" can be pursued.

Interviewing and Recruiting

Communicating vision and metrics while leaving the "how" and some of the "what" to your team should go a long way to generating motivated individuals within your organization. It is also a reality that you don't keep motivated individuals on a team for their entire careers. People need to grow and learn, and this means they will usually spend a certain part of their career with one team and set of technologies and then will move to other teams. This is healthy and should be encouraged whenever possible. To keep new talent flowing into a team as people move on, a team needs to recruit new motivated individuals from outside the organization. There are three phases to recruiting new motivated individuals – first attract, then screen, then interview.

Attracting Motivated Individuals

In order to attract motivated individuals, you need to be able to tell your story to people outside your team in a way that helps attract people who would be motivated to work on your project. Usually you will use job descriptions to communicate your story. When you do this, it is very important to write a job description that not only communicates the vision that your team is trying to achieve but is also extremely honest about what you are expecting from an individual who would be working in your team. There is nothing worse than writing a job description designed to attract a candidate to a job that doesn't really exist as described. Sometimes with data engineering teams, there is a temptation to "overemphasize" the machine learning parts of roles while "underemphasizing" the engineering portions of a role. Even worse, occasionally we have seen machine learning opportunities being used as bait to attract candidates to roles that in reality are 95% traditional engineering roles. An individual who is attracted to a role based on a job description that proves false is not going to be a motivated individual.

It is also important to think carefully about the language you use in a job description to ensure it is inclusive and attracts the broadest range of candidates. There are lots of potential pitfalls that one can fall into when writing a job description that will scare candidates away from your position. The following are some tips:

- Avoid creating large laundry lists of technologies that you expect the candidate to have mastered. A good candidate who has experience with one particular machine learning toolkit can quickly pick up another. Focus instead on broader machine learning techniques that you are sure will need to be used in your problem space – broader terms like NLP or RNNs as opposed to TensorFlow or PyTorch.

- Give some idea of the agile processes and approaches you are using. Candidates like to hear details like "You will be working with a team of five other engineers following Agile principles to make progress on the problem in 2-week planning increments. We believe that everyone should participate in shaping our direction and planning the work we do."

It is also all too easy to insert gender-specific bias into a job description. The National Center for Women and Information Technology (ncwit.org) has provided seven helpful tips for writing job descriptions that attract the widest range of candidates:

1. Avoid superlatives or extreme modifiers, for example, phrases like "rock star" or "world-class." Instead try "truly innovative" or "dedicated and committed to creative problem solving."

2. Avoid gender-specific pronouns (he or she).

3. Make sure that all the "required" qualifications are truly required and try to build in as much flexibility as possible.

4. At the beginning of the job description, include a short but engaging overview of the job.

5. Avoid long bulleted lists of responsibilities or qualifications.

6. Make sure that all pictures and graphics include a diverse range of people.

7. Examine job descriptions for subtle biases in "masculine/ feminine"-associated language.

For more examples, see `www.ncwit.org/jobdescriptionchecklist`.

Screening

Now that people are starting to apply for your job, you will start to get resumes and applications. Once again, at this stage, it is important to not let your biases get in the way to potentially screen out great candidates. It is all too easy to look through a stack of resumes and be biased toward candidates that have similar backgrounds as you. Instead of immediately looking at the school a candidate went to or the last company they worked for, look instead at the projects they have worked on and think through how those projects may prepare or even bring new perspectives to the project you are hiring for.

It is also important at this stage to begin talking face to face with the candidate – we typically like to do a 50-minute interview over Skype where we can ask the candidate to work through some sort of a technical problem that helps us see how a candidate thinks and works through a hard problem. The closer this problem is to real-world things you are working on, the better. Try to work on actual code in this phone screen interview. Don't be religious about the coding language used by the candidate. The candidate should be able to code in the language they are most familiar with, and as an interviewer, it is an opportunity for you to dust off skills in other programming languages that aren't your favorite. In general, we aren't fans of using pseudo-code as the exclusive "language" of the interview – pseudo-code can be a good place for a candidate to start explaining the algorithms they are going to use, but we feel it is important to see if the candidate can also express their ideas clearly in their favorite programming language.

Paired Programming in an Interview   One successful technique we've used in phone interviews is a paired programming approach. As the interviewer, you can write a unit test in a shared coding window (some ideas on how to provide that shared window follow). The interviewee can then "take over the keyboard" and write the code to make the unit test pass. The interviewee can then write the next needed unit test to solve the next stage of the problem. You can then write the code to make the interviewee's unit test pass – and so on. Skype provides a nice shared coding window that can be shared via a browser here: `www.skype.com/en/interviews/`. We have also just opened an instance of Visual Studio Code and used screen sharing to share an IDE during an interview. Visual Studio also has a feature called "Live Share" that makes this easy.

Interviewing

Ideally, you'll also have a final opportunity to do additional interviews with the candidate with other team members. The ideal situation is that within the span of at most a day, a number of employees will interview the candidate on site to determine if they would make a strong team member. It is a challenging task to determine if a candidate will be a motivating and motivated individual on your team or an individual who is demotivating and demotivated. If you reflect on your experience with existing teams, you will find that there are individuals that turned out to be solid contributors who you may have initially thought were below your standard. Conversely, you will also have worked with people who seemed amazing in an interview situation that turned out to be pools of toxicity. How can you perfectly determine in the course of a day where your candidate fits in this spectrum? The truth is that you can't, so you need to be a bit more Zen about the whole process.[1]

Firstly, think of the interviewing process as a classifier. You want to set your threshold in interviews for high precision. This may mean that you will reject candidates that would have been great – but don't sweat it. But be very careful that the "features" that your classifier is using to evaluate fit are ruthlessly pruned of features that reflect your own bias. Reject any features to your classifier that are similarity score based – for example, this person looks like me, thinks like me, went to the same school as me, or likes the same thing as me.

Secondly, keep in mind that beyond very small teams, hiring is a fluid, continuous process. If at all stages there is clarity about the goals and expectations of the team,

---

[1] If you have the opportunity, rather than relying on a limited interview process, having an internship program provides a far better picture of how an individual might perform. This allows you to see firsthand almost every aspect of the candidate's capabilities and potential.

you can optimize the team over time. In other words, you are a business, so you can encourage false positives to find other positions.

With that out of the way, let's explore what to look for in interviews. We generally think in terms of *capabilities* – skills that are already there – and *potential*, skills desirable for your team that the candidate can develop. The former can be determined by direct lines of questioning, whereas the latter requires a little more indirection and skill on the part of the interviewer. Capabilities and potential capture the current and future skills of the candidate; they also capture some idea of the addition to the overall toolset of the team and if they will amplify or complement or both.

The best career advice we've received is to follow your passion – the employment will take care of itself. Passionate individuals tend to self-select for fit in the sense that if their work is not aligned with their passion, they will look for other opportunities. With this in mind, we find it very useful early on in the interview process to understand what the candidate is passionate about. While this may seem like a simple thing to figure out – just ask "What are you passionate about?" – we've found that many if not most candidates don't believe they are actually passionate about anything. Don't believe them. While many won't respond with a highly formed definition of their passions (such as "I want to build the ultimate AI"), many people have hidden passions that show in non-application areas. For example, we've interviewed and worked with people who are passionate about engineering practices, who are passionate about service architectures, and who are passionate about teams and collaboration, particular emerging parts of the science applied to their domain, or are passionate about the company itself. As a manager, your job is to find passion and leverage it for your team. As a hiring manager, your job is to gauge if finding a passion for the candidate you are interviewing is going to be easy or hard. When trying to elicit an understanding of areas of passion, you have to act a little like a good radio interviewer teasing the story out of a reluctant star. Assume that the passion is there, and the effort required to find it is a measure of how easy it is going to be to leverage.

When testing for machine learning or data science capabilities, we generally like to follow a model of escalating questions that mix both direct testing and asking for examples from experience. For example, following our view that metrics are central to development, we like to ask a question about evaluation for a simple problem. For example: "How would you evaluate a classifier for determining if a web page were a home page of a business or not?" This allows you and the candidate to touch on the following:

- The basics of classification, including true positives, true negatives, false positives, false negatives, precision, recall, accuracy, and so on

- Some basic statistics, especially sampling

- The pragmatics of dealing with a very large data set – the Web

From this point, you can extend the line of inquiry into architectural details, especially as they pertain to scale and machine learning specifics if there is an existing capability. When looking for capabilities, you are looking to see if they are a true practitioner rather than someone who has downloaded the latest open source model and run it according to the recipe found on a blog. When looking for potential, you are looking to see if the candidate can recognize common pitfalls – do they recognize assumptions that are being made about the data?

When interviewing for more senior positions, we tend to look for qualities that will work within an agile development framework. Again, when testing capabilities, this can be done by direct questioning. In fact, the agile manifesto principles suggest a number of good lines of discussion: How have you dealt with changing requirement? How have you worked with customers during the development of a project? Where have you invested in technical excellence to support a project? What motivates you?

Career Management of Motivated Individuals

Once an organization successfully creates motivated individuals and recruits new motivated individuals from outside, the next step is to retain and grow those motivated individuals. How do you do this?

Heart Tree Star Chair

An important first step is to make it a priority to talk regularly about what individuals are looking for in their jobs and careers. One way to have this conversation is a model Barbara Grant developed for employee managers at Microsoft called "Heart Tree Star Chair."

The "Heart" part of the conversation with an employee focuses on having a discussion about what they love about their current job and in general what they love doing. Questions you might ask include the following:

- What do you love doing?

- What work in your job currently brings you the most satisfaction?

- In all the prior work you have done, what have you enjoyed working on the most?

- What are the things about your job that you like the least?

The "Tree" part of the conversation is about what an employee would like to do to grow and develop in their position. Questions you might ask include the following:

- What do you see yourself doing 1 year from now? 2 years? 4 years?

- As you see other people in the organization, whom do you aspire to be like in the organization?

- What do you want to learn next?

- What skills do you want to develop next or improve?

- How do you plan to stretch yourself in the next year?

Finally, the "Star" portion of the conversation includes questions about what an employee finds rewarding about their job. You should make it clear that it is assumed that everyone wants to be fairly compensated for the work they do – this is certainly part of the "Star" conversation, and if someone feels they aren't being properly rewarded financially, you should prioritize this. But it is also important to understand other aspects of what motivates an employee. For some, they may be motivated by recognition, others by taking on the hardest problems in the organization, others by being able to impact the most people, and others by being able to work with people they think they will learn the most from. Questions you might ask include the following:

- What makes it worth it for you to continue in this job?

- In addition to monetary compensation, what is most rewarding about working here?

You can optionally add a fourth element to this discussion called "Chair." The Chair question is to determine what keeps them in their current role and what would cause them to leave it. Questions you can ask here include the following:

- Why are you still in your current position?

- What would most likely cause you to decide to leave your current position?

- As you think about your next position, what would it be?

- What is your dream position long term?

Changing Chairs Can Be Good   Remember it is not necessarily a bad thing for someone to change their chair. Often the quickest way someone can grow and progress is by doing something different, even though they are great at what they are currently doing. We have had many experiences where a project seemed totally dependent on a particular person being in a particular role, only to find that when that person left, new people were able to fill the role even better – or at least in a different way that overall created a stronger organization.

From IC to Lead/Architect/Manager

Most reasonable people expect a certain measured progress toward their career goals. It is good for a boss and an employee to be clear with one another what the expectation is for both career goals and the time frame to achieve those career goals. It is good to understand first what path an employee is on. Does the employee want to become a manager in the future or do they want to follow an "IC" path? —Microsoft-speak for an individual contributor with no management responsibilities.

The most common career paths look like this:

- Junior to expert developer: This path is for a person who wants to be hands on in development throughout their career. People in this career path take on more and more challenging projects over time. They may choose to become very deep in a particular area which is valuable – some developers work in a particular area and similar set of applications for their entire careers – while others choose to become knowledgeable about a wide range of techniques across multiple areas. Both types of developers – expert specialists and expert generalists – are valuable to a team. Someone on this path needs to be given assignments that continue to grow and stretch their technical ability.

- The technical leadership path: This path is for a person who naturally takes charge of things technically but doesn't necessarily want to manage people. People in this career are excellent at communicating and refining technical ideas and naturally provide technical leadership to projects they are on. This path ultimately can develop into being the team architect and providing technical direction for larger and larger projects. Someone on this path needs to have opportunity to lead projects that initially are small projects and small teams but over time grow to encompass larger projects and larger teams. People in this career path need to learn to lead by influence and not authority.

- The people management path: This path is for a person who wants to grow and develop high-performing teams and influence and impact other people's careers and lives positively. People in this career path should have a passion for creating environments for motivated individuals to thrive. They should be given opportunities to take on people reports and lead organizational change initiatives to grow into these roles.

- Discipline switch path: Occasionally you will see someone who wants to switch disciplines on a team. For example, we have seen developers who have moved into program manager roles, technical writing roles, and so on. One common challenge on a data engineering team is when someone who primarily has a classical computer science background with little exposure to machine learning wants to fully move to data science roles that are typically filled by people who have more machine learning experience. The way we have handled this is as follows:

- Careful pairing: Typically, on a data engineering team, you will have people who have strong machine learning backgrounds who are driving the most important work on the team. It is useful to spread their knowledge by pairing them with others who are interested in learning and building the same skill sets. But at the same time, you don't want to overwhelm your strong ML people with too many "helpers." Find a balance where you can assign people who want to learn more machine learning on teams where they will rub shoulders with your machine learning experts without slowing down those experts too much.

- Prioritize opportunities: Often everyone on the team from the most junior developer to the most senior developer wants to get more involved in machine learning problems. It isn't always possible for everyone to be involved. There is usually still lots of "non-machine learning" engineering work that has to be completed on a team. Machine learning opportunities can be used to motivate and retain individuals who want to learn. Alternate traditional engineering work assignments with more machine learning assignments so that you continue to make progress on the non-machine learning work while providing opportunities for developers to learn more about machine learning on a regular basis.

Creating a Productive Environment for Motivated Individuals

It is important to monitor the engineering environment of the team to ensure that you are providing an environment that continues to motivate and empower motivated individuals. The team must be mindful of what is sometimes called the "day in the life of a developer." This involves asking questions about a developer's day like the following:

- How much wasted time did you have today?

- How long did it take you to make a change to a model, retrain, and test it?

- How long did it take to make a code change to a feature and get that change checked in?

- How long does it take to verify that a change doesn't break the current product?

- How long does it take to build the product?

- How long does it take to investigate and fix a bug in the product?

Inner and Outer Loop

As you can see, most of these questions have to do with how effectively the developer's time is being used and how much time is being wasted. At Microsoft, teams talk about "the inner loop" and "the outer loop." The inner loop is the amount of time it takes for a developer to change the product on their own machine in some way and test their change. The outer loop is the amount of time it takes once a change has been verified on a developer's own machine to actually ship that change into a production environment for customers to use. It is good for the team to identify what are some of the most frequent things they do in the inner loop and outer loop on the team – for example, retrain models, fix bugs in feature code, or modify a web service – and measure the time it takes. Then the team can relentlessly drive down the time to do those things.

There are many opportunities for improvement that you will find as you start to measure the inner loop and outer loop for your product. Some examples of improvements we have seen to the inner loop include the following:

- An inner loop that included build times of up to 20 minutes was optimized to bring build times down to 2 minutes through refactoring and componentizing the build so everyone only had to build the part of the product they were working on which saved thousands of man hours over a year.

- An inner loop for a data pipeline that could only be verified on a server over a several-hour process was optimized by providing a new flavor of the data pipeline that could be verified in minutes on a local developer machine.

- An outer loop for deploying changes to a web site that had hours of tests was optimized by improving the speed at which the tests ran by parallelizing the tests to be run on multiple machines at once.

How to Be a Hero   If you really want the adoration of your team, do something to improve the inner or outer loop. On one of Eric's teams, the continuous integration system had slowly devolved over time from a 1-hour run to take more than 2 hours to complete verification of any check-in. When a developer stopped his "day job" and decided to really dig in and investigate what was happening, he found a memory leak that was impacting every test that was run. When he fixed it, the integration system was brought back to 1 hour. The team declared a team holiday in his honor.

Tooling, Monitoring, and Documentation

In addition to continually optimizing the Inner and outer loops for your team, you should also think about tooling, monitoring, and documentation. These three investments further improve the development environment, thereby keeping your team motivated.

Tooling: There is nothing better than a tool that makes the developer's life easier. Your team will tell you the parts of the "day in the life of a developer" that are hard and will often have ideas for tools that will make those parts easier. These tools may range from simple scripts that automate repetitive actions to more complex tools that may require much more time to develop. Although you can overinvest in tooling, we much more commonly see that good tooling is neglected with the rush to ship a product. Make time for tooling. Work it into your regular development process in some way. For example, you could choose to do regular tooling sprints, you could choose to rotate developers into sprints where they are working on tooling on a regular basis, or you could have regular designated weeks of the year where the regular product cadence stops to build tooling as described in Chapter 8: Sustainable Development.

Monitoring: There is nothing worse than not being able to figure out why your product is failing in the wild. Often, good logging and good tools to search and query those logs can make the difference between a feature or site failure that lasts minutes and one that lasts for hours. This is discussed further in Chapter 7: Monitoring. A good practice is to have a weekly postmortem meeting about any failures of your product that happened during the previous week. Bring the people into the room who have knowledge of the failure. Do not treat it as an opportunity to lay blame at the feet of the developer who broke the product. Instead, treat it as a useful learning opportunity to figure out what mitigations can be taken to ensure the same kind of issue doesn't recur in the future. Talk about both "time to detect" and "time to resolve." Time to detect should be tracked for major issues to keep the team honest about how quickly they detected an issue impacting customers. Often, time to detect can be reduced by more efficient monitoring. Time to resolve can be reduced through better strategies for updating the product in place to fix an issue. The output of weekly postmortem meetings should be work items that are tracked in subsequent postmortem meetings to ensure the work items get completed. This is discussed further in Chapter 8: Sustainable Development.

Documentation: A final piece that is often lacking in a team to support motivated individuals is good documentation. Some questions you should ask here include the following:

- Does the team have enough documentation in place that a new hire could join the team and get up to speed by just reading those documents?

- Are the most common developer tasks on the team (training a model, updating a feature, deploying a feature) documented somewhere in a step by step way so a new hire can do those tasks?

- Can a developer look at your source code repository and find appropriate readme documents in key directories that explain the contents of those directories?

Investing in documentation is one of the most common tasks that appears in developer surveys at Microsoft. Usually you will have motivated individuals on your team that are good writers and want to invest in documentation. Make time in your schedule and empower those people to write documentation.

DocFX + Git   One system that we have been successful using for documentation is the combination of DocFX and Git. DocFX is a system that can take markdown documents and provide a nice navigation and search experience. It is used to power docs.microsoft.com. When combined with Git, you can keep a change history of the changes made to your documentation and use all the power of Git to manage and maintain the documentation. For more on DocFX, see `http://dotnet.github.io/docfx`.

Developer NSAT

We have already suggested several quantitative metrics to measure the productivity of your environment:

- What is the "inner loop" time for common developer tasks in the inner loop

- What is the "outer loop" time for common developer tasks in the outer loop

- What is the time to detect an issue in the product

- What is the time to resolve an issue in the product

In addition to these quantitative metrics, it is useful to gather a qualitative metric as well. One metric that we have used at Microsoft that is particularly useful to track is known as developer NSAT (net user satisfaction). It is very easy to track developer NSAT for your team. On a regular basis (we do it monthly), ask everyone on your team this question:

How satisfied are you with your overall developer experience and productivity on our team?

1. *Very dissatisfied*

2. *Dissatisfied*

3. *Neither satisfied nor dissatisfied*

4. *Satisfied*

5. *Very satisfied*

You should also ask one additional "fill in the blank" question:

What investment or change would most directly improve your overall developer experience and productivity?

Ask these two questions in an anonymous survey and make sure the team knows it is anonymous. This will encourage honest responses. Then from that survey, you can calculate an NSAT score by this formula:

$$NSAT = 100 + \% \text{ of very satisfied} - (\% \text{ of very dissatisfied} + \% \text{ of dissatisfied})$$

So if you get these results for your survey

Very dissatisfied: 10% of respondents

Dissatisfied: 18% of respondents

Neither satisfied nor dissatisfied: 31% of respondents

Satisfied: 40% of respondents

Very satisfied: 1% of respondents

your NSAT score would be

$$NSAT = 100 + 1 - (10 + 18) = 73$$

Track NSAT regularly and ensure that it is moving up and not down. Invest in the work suggested by the "fill in the blank" part of the survey to keep NSAT moving up on your team.

Supporting Motivated Individuals Outside Your Organization

One area that teams sometimes forget to leverage is thinking about how to motivate individuals working outside of the team and even outside the company. There are several major ways to leverage these external individuals. They include engaging with the open source community, publishing papers, and providing an extensibility story for your project so motivated individuals can build on it and extend it.

Open Source: Leveraging and contributing to open source is a great way to leverage motivated people outside your company and a way to motivate people inside your team. Developers typically feel good about making a contribution to an open source project as it can outlast a contribution made to an internal project. It also can impact many more people than code only written for a smaller project. Where possible, find ways to contribute to open source projects. Also, by using open source projects, you benefit from motivated individuals outside your company who are continuously improving the open source projects you use.

Publishing Papers: Publish and communicate the areas in machine learning where your team is innovating. Attend machine learning conferences where possible and build relationships with researchers in areas that overlap your team's concerns. Keeping involved in the machine learning community can be very motivating for your team and also gives you opportunities to leverage motivated individuals outside your organization who are working on the next breakthroughs in machine learning. Often, the problem you want to solve on your team will already have multiple published papers that can be consulted to get ideas on how to solve your problem. Search the literature and leverage the research that has been done in your problem space. Keep up to date on new publications impacting your area of interest as they come out.

Extensibility: A final way to motivate individuals outside your team and company is to provide a great third-party extensibility model for your product. This might be as simple as a REST API to allow other parties to call into key subsystems of your product and leverage your results in their applications. Think about all the problems that you thought of solving in your product, but you aren't going to get solved because of limited development resources and time. Then consider whether you could expose APIs that would allow third parties to leverage your product to solve those problems.

Conclusion

In Chapter 5, "Motivated Individuals", we have discussed how you can build projects around motivated individuals. We discussed the importance of rewriting frequently. We described some leadership ideas around how to set a vision and metric targets while trusting motivated individuals to figure out the best way to achieve the vision and metrics. We discussed how to find and hire motivated individuals and how to retain and grow their careers once they join. We discussed the importance of having a productive development environment and measuring inner loop and outer loop times for common tasks as well as developer NSAT. Finally, we talked about how to leverage motivated individuals outside your team (which also will motivate people within your team).

In Chapter 6, "Effective Communication", we will describe techniques and strategies to facilitate communication around data.

CHAPTER 6

Effective Communication

*The most **efficient** and **effective** method of conveying information to and within a development team is **face-to-face conversation**.*

—agilemanifesto.org/principles

There is no shortage of topics to discuss in a data engineering team: *We looked at a sample of the input data, and here's what we found. Is this surprising? Currently, the classifier is predicting false positives for these examples. Does the customer care? We compared the latest results with those at the beginning of the month and found this weird bias. What do you think? Do these examples fit in the current definition, or should we update the judgment guidelines? Our labeling team seems to be constantly disagreeing on these cases. Are they important? We think it's okay to get the predictions wrong in these cases. What do you think? We believe this is the right metric to capture overall progress. Do you agree? We reviewed a sample of the data, and it looks like the object model to capture all cases looks like this. Is it too complex?*

If you are attempting to deliver any remotely interesting inference over a large and varied data set, there's going to be an almost constant stream of questions, insights, partial results, and ambiguities to resolve and decisions to be made. What is the best way to answer these questions, read the tea leaves of intermediate results so that you can continue to make progress, keep the team in sync regarding the current thinking about the model that you are building of the world, and unblock individual developers as they navigate the data science search space? The modern development workplace offers a plethora of modes of communication both digital (email, IM, video chat, group chats, code reviews, digital scrum boards) and corporeal (face-to-face communication) – what are the best communication choices to make in the agile data engineering project?

Principle 6 of the agile manifesto advocates face-to-face communication. At the time it was written (2001), the alternatives available to development teams were limited when compared with the variety of communication platforms enabled by the Internet.

© Eric Carter, Matthew Hurst 2019
E. Carter and M. Hurst, *Agile Machine Learning*, https://doi.org/10.1007/978-1-4842-5107-2_6

You can think of the sixth principle as really saying it's better to talk face to face than it is to start up email threads. So what does face-to-face communication have to offer that email lacks?

- Face-to-face communication is **immediate** – when talking face to face, ambiguities, misunderstandings, and confusion can be identified in real time and corrected right away. Asynchronous communication (email, IM, etc.) is at-will, and the latency is a function of the priorities of the individuals involved. Often these priorities are mismatched – there is one individual for whom the topic of the communication is of higher priority than the others.

- Face-to-face communication is **human** – we can take full advantage of the richness of human discourse including nonverbal communication (I can see that you don't seem to understand what I'm saying). All other forms, including voice and video chat, do not have this quality – and no, emojis don't make up for this deficit :-(

- Face-to-face communication is a **shared experience** – we were all there, we saw the same thing, and we agreed to some outcome. A contract witnessed by the team is very valuable in ensuring accountability (though as we will see later, it is not always sufficient).

- Face-to-face communication is **rich and interactive** – I can talk about a data set in a rich way, interacting with the data content both physically (pointing to data graphics that are projected) and computationally (using an application to transform and interrogate the data). I can also draw things on a whiteboard and modify that drawing as I learn from others what is unclear about the emerging picture.

All of this adds up to saying that face-to-face communication, in the context of an engineering project, is the most *efficient* form of communication. With these factors in mind, we can look at the options for digital communication in the modern engineering workspace in terms of *immediacy, richness,* and *shared experiences*:

- Group messaging systems such as Slack and Microsoft Teams offer *potential* immediacy (real-time chatting – though an immediate response is not enforced by the platform). This can be frustrating to the instigator if they are blocked by the issue being discussed.

They then have the additional burden of wondering "Would it be rude to ping?" "Should I drop by their office anyway?" or "Are they even in the office?" – a similar degree of richness as one would find with email, which is to say very little – and they lack the essential interactive nature of face-to-face discussion. You can't really interact with content in a shared way. In addition, many of these systems suffer from a threading or interlacing problem. Communication channels can mix and interlace different threads between different combinations of individuals. Recent improvements to systems like Teams and Slack have attempted to address this specific issue by introducing the ability to respond to a specific message, but have not yet arrived at an ideal solution. At the time of writing, the solutions on both platforms introduce some cognitive overhead and are prone to user error resulting in thread fragmentation.

- Video chat systems offer much of the richness of face-to-face communication (participants can read many cues from the video of participants, though not all, making it more of a human interaction), and through content sharing mechanisms, they offer the rich interactions with data. The limitations of video conferencing – a slight degradation in the ability to read all the nonverbal cues of participants, especially as it pertains to turn taking in discourse – can lead to a slightly stilted but therefore more structured interaction. We've known some executives to require meetings that could be done face to face for most of the team but, with some required attendees remote, to be done online to make the interactions equitable for all participants (and not bias interactions to those who can be present physically).

- Modern code review systems (such as Microsoft's Azure DevOps) provide asynchronous, shared, and appropriately rich experiences around code. Reviewing code can track quite complex interactions between authors, reviewers, and automated components of the system – build pipelines, testing pipelines, and so on. Platforms are becoming sophisticated enough to integrate chat channels into any content system where identity can be used. Similarly, digital scrum boards facilitate many forms of interaction. They are essentially the object of interactive discussions around planning.

One big advantage of digitally mediated interaction is the built-in capability to capture and add value to the content. For example, using a video conferencing platform to host brownbags, scrum team demo meetings, and so on means that the event can be recorded. The recording can become a sharable asset for the team. In addition, some value can be added in both real time and offline – recent advances in machine translation, for example, mean that spoken language can be translated in real time during a meeting. In addition, digital platforms mean that APIs can be opened up to allow bots (intelligent agents capable of multistep interactions with people and content) to contribute to any process. For example, a bot can be built to keep an eye on a code base to look for certain conditions, report them to the engineers involved, and mediate multistep resolutions with those individuals.

Traditional engineering projects involve discussion around systems (architecture, platforms, database schema), processes (planning, execution), operations (deployment, monitoring), and experiences (user interfaces and information architecture). The sixth principle advocates face-to-face communication to resolve issues on these topics. Data engineering projects add to this group interactions around data and inference algorithms. As we will see, the types of discussions to be had and the tools to effectively support those discussions in an interactive manner are distinct in many ways from the traditional engineering context. We will talk about some specific activities that data projects require and how these can be facilitated in face-to-face meetings as well as how discussions around data in meetings found in the traditional agile workflow can be best supported.

The Ephemeral Nature of Face-to-Face Communication   While face-to-face communication has many benefits and teams should use it as much as possible in day to day interactions, it has one fundamental problem which must be confronted in organizations of even a modest size. Face-to-face communication is ephemeral – there is no built-in record of decisions, agreements, designs, alternatives, hypotheses, and so on. In our experience, the immediacy and pragmatism of face-to-face communication needs to be supported and enhanced by a number of best practices that produce some form of digital artifact. This is particularly important in large organizations.

Building a culture that can efficiently combine face-to-face communication with appropriate and timely creation of supporting digital artifacts (both code and non-code) is a challenge that should not be underestimated. Many developers consider the creation of code as the only "work" that they signed up for and any additional tasks involving documentation, planning, and so on as unnecessary overhead. On top of this, if the selection of a content platform and team process is not well thought-out, the team can experience the real overhead of refactoring documents within and across content platforms.

Let's complement our description of the benefits of face-to-face communication with the features of content management systems.

- Cost of creating and editing content: If a content management system makes it easy to create and modify content, it means that the team can very quickly get started. Systems like Microsoft OneNote allow users to get up and running in a matter of minutes. On the downside, unconstrained content creation can very rapidly lead to messy, unorganized content. Pages and sections are forgotten about and become stale. As the cost of content creation increases, it can lead to the team making more considered, better decisions about how to organize and create content. On the downside, a high cost of maintenance can lead to the team putting off and ultimately abandoning the creation and upkeep of documentation.

- Search capabilities: The less constrained and more open a content system is, the more search becomes an important feature. Poorly structured content that can't be searched becomes almost useless. Microsoft's OneNote, for example, while providing a low cost for creation suffers from mediocre search functionality in our opinion.

Platforms that are commonly considered for managing team content include the following:

- OneNote: Microsoft's WYSIWYG wiki-like system. It has a very low barrier to entry – just start typing – but can lead to large, unstructured, and difficult-to-search content.

- Checked-in documents to the code management system (e.g., Azure DevOps, GitHub): Checking in content allows the team to bring some of the discipline of code reviewing and even testing to the document space. A human-readable, text format should be chosen (markdown, good; pdf, bad) that works well with the built-in comparison tools used for code. If you can establish this culture, then maintaining things is not too cumbersome – you need to add "review document changes" to your workflow when reviewing code.

- Standalone wiki: Systems like the hugely popular MediaWiki platform should be familiar to (or easily adopted by) your team. They offer more constraint than OneNote, and editing requires more intention (which can be a good thing). You can also consider DocFX and Git if you need a workflow that more approximates the workflow for code (see github.com/dotnet/docfx). Atlassian's Confluence is another good choice.

- Word + file system or SharePoint: For important definitional content (such as specifications or requirements) where the content doesn't change frequently, the formality of a full-blown word processor can be a useful tool. Such documents do best in a well-structured document space.

The challenges around non-code artifacts are very human – they relate to discipline, behavior, and habit. It is very hard to come up with an analogy for building and testing documentation. One general approach to help with success in this area is to limit the types of documents that can be created. With this scoped and well-defined set of artifacts, it is far easier to say each check-in is required to include comments for public methods or demo meetings have to be run directly from the experiment log. Some of the artifact types that we have worked with include the following:

- Experiment logs: The data science part of our work is, well, a science. And like most scientific endeavors, data scientists explore various spaces to discover where the good stuff is. Exploring these spaces generally involves a data set, an algorithm, some parameters, some output, and an analysis. There is great value to the team as a whole if records of these experiments are created, made available, and searchable. Even better is if this form of documentation can link to persistent (and immutable) versions of the data involved – to go even further, if some form of serialized version of the experiment can be referenced so that others can even rerun the experiment or inspect all aspects of it (versions of code, parameters, etc.). As the manager

of a team in this space, a deep investment in these capabilities
(either researching third-party systems or building tool chains
internally) is an important commitment. Just as we take advantage
of powerful systems for enabling and integrating unit testing for
our code, persistent experiment description and reproduction will
pay dividends to the team. A very popular platform for this is the
Jupyter Notebook.[1] Jupyter documents interlace cells of text for
human consumption with cells of code which share context for the
entire document and which can be run, manipulated, and expanded
interactively by anyone with access.

- Schema documentation: Deriving schema for anything but the most
  trivial cases requires a reasonably deep study of the domain. We
  experienced this when modeling the expressions used on web sites
  by businesses to describe the business hours of local entities and
  when designing an abstract document representation that could
  be used to describe web pages, pdf documents, SharePoint pages,
  and PowerPoint documents. Capturing the examples that motivate
  everything from the high-level structure of the schema and the
  inclusion, or deliberate exclusion, of certain cases will help the team
  determine what to implement and how.

- Judgment guidelines: Very much related to schema documentation,
  judgment guidelines are, in some sense, the most concrete "truth" for
  the team as they are used to derive the metrics that guide the team's
  work and the impact (and reward!). Judgment guidelines are the
  interface between the rich model of the world that you are attempting
  to infer for your raw data and the high-volume data labelers that you
  will likely be employing to generate training and evaluation data. As
  you can imagine, it is important to get these right. We've developed a
  number of best practices around judgment guidelines. They should
  include standardized descriptions of how to use the specific tools
  for the task. Many tasks will involve annotations of some sort, and
  these should be driven by a common platform. They should include
  an intuitive description of the concept being captured. They should

---

[1] https://jupyter.org

describe specific cases for exclusion and inclusion. They should all be backed up by concrete examples showing exactly how the examples look in the tool and the state of the tool before and after the label has been applied. Finally, the document should be strongly versioned – using both references to prior versions and a description of what changes have been made since the last versions.

As the preceding text suggests, there is no magic bullet when it comes to content. This is where the human factor plays an important role. As a manager, can you encourage or require the team to keep abreast of content tasks?

Discussion Around Data Is Necessarily Interactive

Different team members, and members of other teams, will have different views of the data sets involved in the project. They will see different aspects of the things being modeled in the data; they will have different views of how to evaluate the quality of the representation or inference results; they will have different views of where the risks are. You will also have team members who are interested in the data with respect to the ML tool chain and the production architecture. Bringing these different lenses to the discourse around the data can only benefit the team.

In a recent project to improve address extraction from web sites, while our new extractor performed better in our well-constructed evaluation sets, members of the team responsible for chain businesses, and large, complex entities such as museums and hospitals, observed that there was actually a regression on their subset of the data. If we were just to use the existing metrics and not work across the team, we would have missed this insight. In another setting, a partner team questioned our use of a per document average for precision and recall, suggesting rather that we use global precision and recall for the classifier we were developing. These different approaches gave quite different numbers and helped us look at the data in a different way.

When constructing a new schema to represent the business hours of local entities, an in-depth study of the breadth of expressions revealed that the schema would have to be quite sophisticated, having to represent rules around holidays, repeated patterns of opening and closing, and nonspecific times like "sunset" and "when the last customer leaves." The team responsible for implementing the schema came up with a solution that required tools to inspect and understand instantiations. However, when the result of this research into the representation was shared with downstream consumer teams, there was considerable pushback on the complexity (which could contribute to additional errors in inference) and the form of representation (a simple, textual serialization was preferred, which would allow data engineers to very quickly read and understand instance of the data when debugging and evaluating the quality of the data).

In another project whose goal was to extract the main content from web pages – stripping out the navigational, banner, and other non-primary text – we found that our metric was biased to small documents (for which there is less room for error when identifying the specific components that make up the rendered page). This led to an important conversation with the consumer of the data in which it was established that these smaller documents were actually not of particular value and so constructing a classifier to identify and remove them was the way to go.

Individuals will also bring different assumptions and biases. Again, by combining the views of team members, the team can continually correct and improve their capabilities. Discussing data, inference system design, and the results of implementations will also provide opportunities for the more experienced data engineers to mentor junior team members.

We found that in practice, the only way to have a conversation motivated by differing and often conflicting views and theories is to have the data set at hand and the ability to interactively answer questions. Any statement about the data that is not supported by analysis is a hypothesis which the data will either support or refute. It is very inefficient to answer each question with a scheduled work item followed by a discussion. Rather, as often as possible, interactive tools should be part of the discussion and used to directly select, filter, and transform data sets (or samples).

Data Tool Basics

To support discussions centered on interactions with data sets, it is important to think about best practices including requirements for tools, sampling methods, and data presentation. Remember agile practices improve when we can bring efficiencies, even micro-efficiencies, to any part of the process. It is important, however, to remember that efficiencies have to be at the team level and not greedily applied to individuals. For example, if you are presenting data to your team, putting in extra effort before the discussion so that the meeting will be as productive as possible is a win. The presenter has had to spend more individual time to produce a net win at the team level.

Requirements for Data Discussion Tools

There are many options for tools that will support data discussions. Perhaps the most obvious is the humble spreadsheet (Excel, Google Sheets, etc.). You will certainly consider implementing in-house tools when you come up against the limitations that the generality of these tools impose on your analytical scenarios. The basic requirements for these tools are as follows:

- Loading

    - An obvious requirement, but be aware of the tool's performance when asked to load large data sets. In addition, some tools, while permitting the loading of large data sets, fail to then perform when processing that data in any meaningful way. Good tools will at least implement data and UI virtualization; better tools will be able to subsample data on the fly to provide the user with a representative view of the larger data set.

- Transforming

    - To subset data through filtering and the application of additional predicates. Once you have the data up, you are going to want to filter it to look at different subsets on demand and potentially apply arbitrary filters requiring some sort of method invocation.

    - To generate samples of data interactively for quick, in-place evaluation. In some cases, like spreadsheets, sampling can be done through a relatively simple workflow involving the addition of a column of random numbers that can then be sorted to shuffle the rows in the data. A more sophisticated system would provide this capability in a built-in manner.

    - To compute additional data derived from the source data. For example, you may want to add a column to your table which indicates if the value in another column is above or below a certain threshold. This pattern of building up analytics through derivatives is very powerful and one that is a basic feature of the standard spreadsheet application.

- Aggregation

  - To perform various aggregation functions such as summing, averaging, computing distributions, and so on. Again, basic spreadsheets support this type of operation, but only through a workflow of sub-steps.

- Visualization

  - To visualize the object data that exists at various stages of your system including inputs and outputs.

  - To subset the factors in the data to allow for easier viewing, for example, removing unwanted columns from a table.

  - Charting to show trends, distributions, and relationships.

Making Quick Evaluations

The recipe for making quick ad hoc evaluations using a spreadsheet or similar table-oriented data tool is as follows: firstly, associate each data item with a random number; then, sort the data by this random number. In some sense, any contiguous subsequence of the data can be considered a random sample. However, when proceeding through the data to make judgments, it is important to determine ahead of time how many you will look at. If you don't, you may end up completing the evaluation at a time when the results look favorable to you. This also means that if you see a summary of data from a colleague which claims that they looked at a sample of some arbitrary number of examples, you should question why that number. They may have stopped at a point in the data that gave them a nice result, falling for confirmation bias – finding support in the data for a preconceived conclusion. Keep in mind that random data can be clumpy – which is why if you toss a coin, you don't generate a sequence of HTHTHT but rather a sequence with clusters of Hs and clusters of Ts. Another thing to keep in mind is that if there is a large effect in your data, it will probably be exposed by a relatively small – and therefore cheap to compute – sample. For example, if something happens in 10% of your data, the probability of not observing it in a sample of 50 is about 0.5%.

Given a sample, we often want to then make a quick analysis of the factors found in the data. For example, imagine we are evaluating an extractor for a specific named entity operating over web pages. We might list the following reasons for errors:

- Data is in an image (our approach is text based and so can't access the content if it is in image form).

- Data is not in a text node (the entity might be referenced in an attribute node of an element rather than a text node child of an element).

- Content is missing from HTML (the page might use some render-time dynamic process to pull up the content from a server and modify the DOM to allow the user to see the desired information).

However, what we often see is that it is almost impossible to a priori determine such a list – it is only through inspection of the sample that we can really discover the factors involved. Consequently, the process that is most commonly adopted is to use a spreadsheet with one example per line and, as we analyze, create columns to capture the factors of the analysis. After a few examples, you will start seeing repeated factors. If you follow the practice of placing a 1 or 0 in each factor column according to the presence or absence of the factor, then after reviewing the examples, it is trivial to read off the frequencies of the factors in the analysis. Figure 6-1 shows an example of factor analysis over some URLs.

Rand	Url	AddressInImage	AddressMissing
0.117273	http://matt.com	0	0
0.450732	http://baz.com	1	0
0.472747	http://bar.com	0	1
0.491948	http://rohai.com	1	0
0.737318	http://foo.com	1	0
0.519538	http://kusanku.com	0	0
0.608302	http://eric.com	0	1
0.37673	http://heiku.com	0	0

Figure 6-1. *Example of factor analysis. Note that after sorting the rows by their random numbers, the random numbers appear unordered as Excel regenerates them on every view of the sheet.*

Mining for Instances

Quick evaluations over small samples are good for surfacing *types* of things in your data. Once you've a few of these, it is useful to be able to explicitly mine for that type to get a better understanding of its frequency and makeup. For example, by looking at 50 examples, we can be pretty sure that we will observe things going on at or above the 10% frequency. However, we will still be somewhat vague on the true size of this phenomenon. By looking at a few cases, we can build some sort of predicate (or heuristic) that will allow us to directly pull these examples from the set (hence the need for a tool that can support somewhat arbitrary predicate execution).

Sampling Strategies

The more you learn about sampling strategies from experience with data, the more you can look forward to long discussions about which way to sample for a specific data set and a specific scenario. The reason for this, and the answer to any question about the right way to sample, is that there is no best way to sample and there are many benefits to looking at multiple samples. And while there may be no one best way to sample, there are many bad ways to sample, so it is useful to understand a few of the basic options:

- Simple uniform sampling: This is also called a random sample. In spreadsheet applications, it can be generated by assigning a random number to each row in your data and sorting the data by that number. When dealing programmatically with data sets, the Fisher-Yates[2] shuffle algorithm can be applied; or with larger data sets, potentially infinite streams of data, reservoir sampling[3] can be used.

- Weighted sampling: The simple uniform sample answers the question "If I picked a row from the data at random, what is the chance that I will see some particular effect?" However, it is very rare that your data will be experienced in this way. In web search, for example, people browsing web sites do not uniformly sample from the trillions of pages out there. Rather, they willingly or unwillingly consume a very small portion of the Web. Within that portion,

---

[2]https://en.wikipedia.org/wiki/Fisher-Yates_shuffle
[3]https://en.wikipedia.org/wiki/Reservoir_sampling

some pages are viewed more than others. Consequently, if you want to ask the question "If I were to sample from the population of individual viewing events over the space of all users and all web pages, what is the probability of observing the particular analytical effect I'm interested in?" – for example, if the interesting thing you are interested in is the correctness of the address associated with a business listing on a search engine – knowing that users view more listings associated with chain than with non-chain businesses means that you should care more about those cases where errors are found in chain listings rather than the population of businesses as a whole. Weighted sampling can be carried out by effectively repeating rows in the data according to their weight. So if we assign a weight of 2 to business A and 1 to business B, our sampling would be looking at a population of three data points. Clearly, there is a higher chance of pulling an instance of business A when randomly sampling this view of the data than if we just had two rows in the data.

- Stratified sampling: It is often the case that there are meaningful high-level ways in which your data can be subdivided. It is also often the case that these subdivisions can be associated with expected (or suspected) biases. Let's consider the example of extracting prices from web sites offering products for sale. As, at the time of writing, Amazon dominates this space, we can imagine that a sample of these web pages (either uniform or weighted) would have a large number of pages from the Amazon site. In fact, it is possible that you could end up with only pages from the Amazon site. As all pages on the Amazon site use the same approach to presenting prices, then it is likely that an extractor for prices would either fail or succeed on all these pages. While this gives a certain insight into the performance of the extractor, it doesn't help you understand how the extractor performs over the full gamut of ways in which prices can be displayed. Of course, you are going to fix the Amazon example if it isn't up to snuff, but that would effectively leave you blind to non-Amazon ways of presenting prices. Stratified sampling uses some form of stratification to provide a set of subsamples that are then combined to the complete sample. In this case, we might sample at most ten pages per domain. The sampling algorithm, then, would first sample from the space of domains (we may get amazon. com, bestbuy.com, target.com) and then, for each domain, uniformly samples to get ten pages.

Iterative Differencing

The best approach to tracking and improving quality is to invest in a metric – this includes all the requisite data collection, guidelines writing, tool building, and maintenance. However, sometimes the overhead is too high, but some sort of evaluation is required. An approach that we have found works well is to iteratively diff the output of your system and sample the difference. The data that hasn't changed will have no impact on the estimation of the improvement (or regression) that the latest experiment has yielded. By taking a small, uniform sample of the examples that have changed, you can build a table of the types of changes observed: good to good (both results are acceptable), good to bad (we used to get the right inference, but now we don't), bad to good (a win – what used to be an error is now an improvement), and bad to bad (the result is still bad; it's just different). If this approach is taken, then a by-product is actually labeled data, the accumulation of which can be developed into a metric or at least a regression set for testing.

Seeing the Data

This book is not intended to provide an in-depth review or detailed guidance on presenting data visually. We do want to provide some guidance, however, on how to approach this area. Like all of the non-coding areas that we touch on in this book, data visualization is another where intention is vital – being a data scientist, a machine learning expert, or a data engineer requires that you can communicate and convey narratives about data. Consequently, data visualization needs to be recognized as an area where one can be continually learning, improving, and innovating. Remember that agility requires attention to detail and an investment in efficiency. Clear, memorable, and authentic data presentations contribute to this capability. Be aware also that it is easy to lose intention in this area because there are so many tools available already. However, many of these tools provide access to well-used visualizations that have gained popularity prior to the current era of data projects and prior to the growing body of research into the effectiveness of specific approaches to presenting graphic summary of large-scale and high-dimensional data.

Perhaps the most celebrated example that illustrates the complexity of data visualization is the pie-chart. The issue with pie-charts is one of perception and a whole bunch of neuroscience. It turns out that we are not as good at comparing areas (two dimensions) as we are at comparing the length of lines (one dimension). In fact, you can take this as the first rule of data visualization – don't introduce dimensions in the visualization that are not present in the data. This principle also rejects 3D bar graphs, spider or radar charts, and so on. In addition, pie-charts tend to require the use of color – introducing further complexity and overhead for our brains. So instead of this

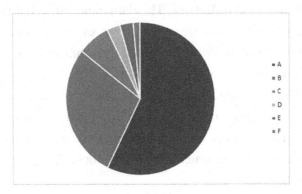

consider doing this

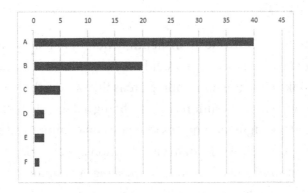

but certainly not this

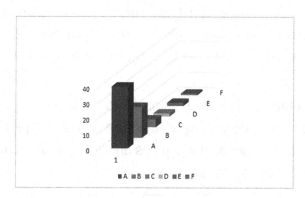

Our advice, then, is to consider data visualization as an integral part of your toolkit. Just as you have invested in understanding how to optimize your code, how to ensure thread safety, how to design for testing, and all other deep arts of engineering, so should you dive into the literature of data visualization and understand how to help consumers of your data analytics understand it accurately and objectively.

Additional Reading   Much has been written about data visualization. Two excellent sources for additional reading are *Envisioning Information* by Edward Tufte and *Information Dashboard Design* by Stephen Few. Also, if you'd like to be inspired by examples of masterful data visualizations, see aka.ms/dataviz2018.

Running an Effective Meeting Is a Skill

Face-to-face meetings are not a simple matter of fixing a time and place. As developers, we should be very protective of our dedicated time for focusing on tasks, and as managers we should be aware of the cost of getting people together (a 1-hour meeting for a modest size group is roughly the same cost as a person-day of work). At a bare minimum, a meeting should have a clear goal and a declared process for getting to that goal. It is often useful to set some expectations regarding individual preparation for meetings (e.g., reading requirements documents, doing individual analysis on a data set, etc.). Those charged with running the meeting should also be accountable for the technical success of the meeting – the shorter the meeting, the more risk there is from techno-fumbles such as projectors that don't work, laptop batteries that need charging, and glitchy video conference integration. In addition, the social side of the meeting can always be finessed – ensure that key stakeholders can make it on time, finding out who has meetings before and after and if those meetings are organizationally of higher priority than yours. Finally, it is always valuable to think through the data required to answer key questions and to ensure that the data is readily available.

Laptops or No Laptops   It can also be useful to set some expectations for engagement in a meeting. Do you expect the full attention from everyone throughout the meeting? If so, then ensuring that laptops are closed and no one is casually reading their news feed is a good idea.

Moderated Meetings

One useful technique you should consider for larger meetings is the moderated meeting format. This format puts some more formality around the meeting and helps everyone participate more fairly in the meeting. One person is designated as a moderator. That moderator has a list of topics that the meeting is going to cover and a time limit for each topic. A primary presenter for the topic is designated to go first. But instead of allowing random interruptions of the presenter, people with comments or questions raise their hand. The moderator puts those people into a "first in, first out" list that is either projected to the screen or kept on a whiteboard in the room so everyone can see the order of the queue. When the primary speaker is ready to yield the floor, people speak in their ordered turn until they are done. At any time, additional people can raise their hands to be added to the queue. The moderator watches the clock and gives time warnings when the topic time is about to run out.

The moderated meeting format has many advantages including the following:

- It puts people who are participating remotely on an equal playing field.

- It makes it easier for people to contribute who feel uncomfortable trying to wrest the floor away from someone else who is talking.

- It helps people who have a tendency to talk too much to listen more to what others are saying.

For more ideas on how to run an effective moderated meeting, see Chelsea Troy's blog post at https://aka.ms/moderatedmeetings.

Pair and Parallel Labeling

Pair coding is an approach to coding in which two developers work together in a fluid and interactive manner writing and testing code. One of the developers leads, hitting the keys, and through discussion and iteration the code is generated. Many errors are almost immediately avoided due to the second pair of eyes present in the process, and the code review effectively happens as the code and tests are written. When embraced, this process can be very rewarding – an efficient way to write quality code and a great way for both developers to learn from each other.

We found similar efficiencies in labeling training and evaluation data in pairs. When labeling web data for a new address extractor, we found that pair labeling had the following benefits:

1. Locating the label targets on the page: Pages can be complex, and sometimes the address isn't always easily located.

2. Refining the address model: As we were not only labeling the full span of the address, but the sub-parts, our discussion during labeling led to refinements in how we expressed the overall address model in our labels.

3. Capturing issues: As one of us is labeling, as a pair we can find and discuss patterns of issues we see and come up with ways to describe and record them.

4. Shared accountability: A danger in solo work, especially when you are tasked with delivering the extractor or classifier, is that you may be lenient when labeling evaluation data. By having another person present, some degree of balance can emerge through the interaction.

We often find situations in which we generally agree at some high level on the definition and role of a concept but find that precisely defining it is tricky. For example, in a recent project, we wanted to provide a primary topic to a set of news articles so that they could be served to users in a content recommendation system. We generated a large set of topics by mining the tags found on popular community sites and went about building a data set to be labeled in a mechanical turk-like system. However, when we started looking in detail at the topics, we started doubting our intuitions about labels.

To better understand where the issues were and how to resolve them, we conducted a number of group labeling sessions. The developer team charged with delivering the topic classification system sat together with a labeling task. The group paired off in teams responsible for ten topics each and labeled a shared document set individually. Whenever a problematic example appeared, the team would discuss the article, figure out a resolution (using the shared context of the labeling thus far), and, if appropriate, update the judgment guidelines adding additional detail and refactoring existing text to reflect the latest thinking.

When appropriate, the teams would discuss issues across the topic. This approach means that many definitional issues can be addressed in real time. The alternative is for one person or group to carry out an iteration of labeling, summarize issues, report back to the team, and then run another iteration.

Data Wallows

We've talked a lot in this book about the central role that metrics take in the execution of data projects. They act as a proxy for the customer that the developers can use in every experiment and iteration of the project, and they act as a forcing function for a number of design decisions, most importantly the schema or data structures used to capture the output of the system. However, there are many moments in the lifetime of a project when a decision needs to be made where either there is no metric or the metric is too general or otherwise insufficient to provide a clean black and white answer. In such situations, the team needs to make some sort of judgment call. Precisely how that decision is made is up to the team, but we want to introduce a general team-based process that provides an efficient and well-informed framework: the data wallow.

Beyond the charming image of data-pigs splashing around in data-mud and having a jolly old time, a data wallow is a meeting in which the team makes decisions based on getting their hands into the data, interacting with it, and asking questions of it. The basic workflow is as follows.

First, the developer facilitating the data wallow (who is primarily responsible for getting the team to arrive at a decision) provides the team with a description of the decisions they are trying to arrive at, the necessary data sets involved, and the tools to interact with the data and, if required, to make inferences on the data.

Second, the team spends some individual time reviewing the data, forming their own opinions regarding the data and the decision to be made and any questions they would like to raise.

Third, the data wallow meeting takes place. The facilitator provides context, a review of the data, and the team contributes their thoughts and questions. During the meeting, questions are answered through inspection of the data – hence the importance of interactive tools, running inferences, and so on.

Let's consider a concrete example end to end and then explore the types of decisions that are typically carried out through a data wallow.

The team had been working on a project to extract business hours from web pages. The project was precision focused, meaning that the extraction technology implemented rules which only generated results if particular constraints were met. In particular, business hours were delivered only if the system found a description of opening hours for every day of the week. The result of this approach to delivering data was that the data quality was very high, but it had limited recall. We wanted to move the needle on recall and so ran a number of experiments using different, more relaxed constraints.

To make the decision regarding shipping the data, the developer organized a data wallow. To prepare for the wallow, he did some evaluation of the results and prepared a deck which captured the evaluation. In addition, he developed an interactive tool that would allow the team to see the results of extraction in real time during the meeting in a format which showed exactly where in the raw document the extraction came from.

The meeting started with a review of the evaluation method. Firstly, a differential analysis was done comparing the data extracted under the new constraints with the data currently extracted for the test sites in production. From 57, 000 sites, the net gain was 3, 300 sites with hours extracted, 3, 500 were new extractions, 200 were regressions (meaning we had extracted them, but the new system now failed to extract), and 200 were modifications (meaning we still extracted but we got different results). The dev projected that we would expect an increase of about 50, 000 extractions over the entire corpus and, from a manually evaluated sample of 100 examples, the precision would be about 90% - slightly lower than the current production performance.

This first slide allows the team to understand the impact of the work and the details of evaluation. Any criticism of the evaluation can be voiced directly in the meeting, and a group decision can be made as to whether the evaluation methodology was appropriate and if the results support a decision to ship.

As the team felt that the evaluation looked reasonable, the meeting then shifted to looking interactively at extraction results for specific sites. Beyond the random sampling done to determine the precision of the extraction, individuals on the team can get an immediate look at how well the system performs on parts of the problem domain that they are individually working on. There were some regressions mentioned – was there any bias toward a specific area? Some sites showed neither wins nor loses but changes to extraction – what did they look like and were they positive, negative, or neutral? Because the developer was armed with examples of each type of result (win, loss, change), the team could view and discuss these cases. Because the developer had an interactive tool at hand to run extractions, the team could look at the results for sites individuals were interested in.

The result of the meeting was that the team supported shipping the changes.

Demo Meetings

At the end of a sprint, we have a demo meeting as introduced in Chapter 4: Aligning with the Business. The dev team, or group of teams if working on a larger, multi-scrum project, gets together to show their work. In formal scrum processes, the team demonstrates working software to the customer. One modification to the normal scrum process of the individual scrum team doing the demo meeting, we found that having a single demo meeting that spanned multiple teams that worked on the same overall project, but in a modular fashion, was a useful approach. While there is something of an additional cost in terms of the length of the meeting, by having teams that worked and depended on each other present together, the work, results, and planning would in a way self-regulate across the broader goals of the combined teams' project.

In data projects, in addition to working software, presenting data analyses, important insights, and the reasoning behind major planning and ship decisions is also an effective use of this time. With or without a formal customer present, the meeting is a great opportunity for the team to bookmark a sprint's worth of work (a celebration, if you will, potentially involving food and beverages) and to report up to direct managers and sideways to peer managers.

Like many events in scrum, the meeting is time boxed. Each scrum team is allocated a specific amount of time (generally proportional to their size – a five-person team might be allocated 10 to 15 minutes). We found that adopting a template for the first, summary slide presented by each team was an effective way to keep the mission of the team front and center and to provide a regular check-in with the key metrics for the team. The summary slide contains a table of the stories that were planned in the sprint, if the story was completed or not, and a table showing the key metrics including the target value, the current value, and an indication of improvement over time. Figure 6-2 shows an illustrative example of a sprint summary slide.

SMILE Infrastructure		Metrics Gains	Achieved this Sprint	Next Sprint Goal
		Q	+.1	+.1
		LDCG	+.1	+.1

Sprint Goal	Met?
Design solution for dynamic web crawling	N
Operationalize delivery of chains with flat directory page	Y
Design site coherence (for knowing whether we can publish as definitive)	Y
Improve annotator robustness by making one annotator run at a time.	Y

Story ID	Story Description	Done?	In Demo?
<ID>	<Description>	<Y/N>	<Y/N>

Figure 6-2. Example sprint summary slide

Compared with activities like data wallows and group labeling, the sprint demo meeting tends to be more constrained. There isn't as much time for open-ended discussion as each team has a limited time slot and, as we've observed, teams get protective of their time (and rightly so). But this constraint is a good thing – one of the key skills in presenting is to aim for clarity and remove opportunities for misunderstanding while being as transparent and as objective as possible. It's not great to burn time in presentations responding to audience questions leading from confusion around the axes of a chart or the meaning and significance of an insight that a data visualization is intended to convey. In other words, the value of getting together in this meeting is the unlooked for insights that come from different people's views of the results, and the more presenters can facilitate this through principled and clear data presentations, the better. Whenever serendipity or confusion strikes, given the timeboxing of the event, we found it a good practice for the scrum masters involved to schedule a follow-up meeting – this allows the overall flow of the demo meeting to continue while harvesting value from the interaction.

Here are some notes on best practices for presentations:

- The audience can read[4]: Once you put something up on the screen, like it or not, people are going to read it. They are going to read it faster than you can present it. You should expect them to read it so quickly that they won't be listening to your explanation of why it doesn't mean what it appears to mean. The only things you can control are the sequence of presentation – when and how much text is thrown up on the screen – and the clarity of the content. If clarity cannot be achieved reasonably in the slide, force the audience to pay attention by sequencing the reveal of information in a way that allows you to prepare them.

- Label your axes: We learned this in elementary school, but it didn't stick. One of the reasons that we get sloppy at this is that some popular data manipulation and presentation applications don't always make it trivial or obvious how to do this.

- Start with the win, then show how you got there: Presentations should be structured as if they could be interrupted by a fire alarm at any moment (similarly, documents should be structured as if the reader might be distracted at any time). If all you achieved was to get through the first slide in the deck, make it count – it should have the high-level summary front and center. Once you have people clear on what you are claiming, you can then take them on the journey of how you got there. The alternative is a meandering walk through some of how you got there interrupted by a request to skip to the results when you realize you are running out of time.

- Prepare for multi-use: You've built a deck with some nice animation and a tantalizing mix of content on the screen and presenter theatrics only to find that the deck was printed on paper and distributed to VPs before the meeting. Now they are confused. At least, be aware if the deck is intended for consumption as a document to be read and design accordingly – our demo meeting decks would be read over later so that the team leader could create a monthly update on team progress for reporting up the chain.

---

[4]Attributed to Nathan Myhrvold.

Don't Forget Testing for Color   Matt's heart sank when he brought up his nicely designed slides with an awesome color pallet only to discover that what appeared with balanced contrast on his laptop screen looked washed out and almost illegible when presented. Also remember that many people (1 in 12 men and 1 in 200 women) have red-green color blindness. Color-blind people can perceive brightness shifts, so make sure to vary the brightness of colors and textures for different values to make your presentation accessible to them. Wearecolorblind.com gives additional tips for making presentations accessible to color-blind users.

Conclusion

In Chapter 6, "Effective Communication", we illustrated and contrasted the principle with a dive into many of the types of interactions that you will encounter as an agile data engineering team. The main message here is that all of this communication is a skill – just as writing code and going deep on specific machine learning methods. Like any skill, it can't be learned, maintained, or improved unless it is acknowledged and becomes an intentional part of your personal work and the team's culture.

Chapter 7, "Monitoring", will discuss how telemetry and mining of telemetry logs can provide important data about whether your product is really working as expected, why and when the product is malfunctioning, and if you are truly providing a good experience to your customer.

Monitoring

Working software is the primary measure of progress.

— agilemanifesto.org/principles

In Chapter 3: Continuous Delivery, we talked about techniques like continuous integration and continuous deployment that ensure that new working software is being delivered not just frequently but continuously. The authors of the agile manifesto clearly point out in this principle that having actual working software is much more valuable than say having a lot of documents that describe some software system that still hasn't been integrated enough to be running. So, in any software project, getting something running quickly and then keeping it running while incrementally adding feature is essential. Some techniques to doing this were described in Chapter 1: Early Delivery.

But we believe that having working software, while critical, is not sufficient to measure your progress. If you don't truly know what your working software is doing while working – if you haven't built measurement into your working software – having working software can be equivalent to the proverbial tree that falls in the forest with no one around to hear it fall. It is likely the software is falling in ways that no one is hearing.

Monitoring Working Software

As part of the development process for shipping working software, you should have a plan for the telemetry you want to collect as part of that working software. That telemetry can be used to not only ensure the software is working as promised for your users but also guide development of future features and measure whether what you have shipped truly meets the needs of the business and customers. There are four questions you should strive to answer with your telemetry. First, is the software really working correctly as designed or is it failing unexpectedly? Second, how correct is the data you are

E. Carter and M. Hurst, *Agile Machine Learning*, https://doi.org/10.1007/978-1-4842-5107-2_7

displaying to users? Third, what are the business goals you are trying to meet, and can you measure that those goals are actually being met? Finally, what needs does your user have, and are you measuring the fulfillment of those needs?

Monitoring and Privacy   We propose that working software can be monitored in a way that protects the privacy of your users. Increasingly, customers demand this as reflected in laws and regulations being passed around the world such as the General Data Protection Regulation or GDPR. We will discuss mechanisms in this chapter that can be used to monitor while respecting privacy of users. However, it is likely in the future that even this "privacy-aware" monitoring may need to be opted into by your users in certain jurisdictions.

An Example System: Time to Leave

As an example to motivate the discussion in this chapter, we briefly describe a system that Eric worked on while working for the Cortana team. The feature was called "Time to Leave" (TTL) and was integrated into the Outlook mobile client on the iPhone. There are similar features available in Apple Mail and Google Inbox. The feature works like this: if you schedule a meeting on your calendar and add a location to the meeting that can be located on a map and if you share location data with the Outlook mobile client, then TTL calculates the driving directions and length of the drive between where you currently are and your next meeting and alerts you on your phone at the correct Time to Leave so you can travel to your meeting and arrive on time. TTL does this based on the current traffic conditions and expected drive time between your current location and the location of the next meeting. So if you had a meeting scheduled at 3 that was 20 minutes away in current traffic, TTL would alert you at 2:30 that you better start driving to your next meeting to get there on time and show you the current traffic map to your meeting. If the meeting was close by and didn't require significant travel time, TTL would just give you the standard upcoming meeting alert without traffic information.

This system had some code in the Outlook mobile client, but most of the business logic and code ran on the server as a set of microservices. It interacted with calendar data stored on a server to get information about calendar changes that might include locations to track. It interacted with Bing's traffic and routing system to get information about current traffic conditions for an upcoming meeting. And it interacted with notification services on the iPhone to trigger the alert at the right time. The iPhone app also interacted with the microservices to turn the feature on or off when the user activated or deactivated the feature within their Outlook mobile client.

Monitoring is essential for any application to fully understand whether it is working as expected, and in this project, it was particularly critical as there were so many moving pieces to the system. Also, many logical operations occurred across several execution surfaces – for example, to successfully turn the feature on, the following needed to occur successfully: code had to run on the Outlook mobile client on the phone, the user had to agree to share location data to the application, code on the server had to run to turn the feature on, additional code on the server had to be kicked off to scan upcoming calendar events for locations, and code had to begin running to listen for future changes to meetings.

Activity-Based Monitoring

The Time to Leave system had tens of thousands of lines of code in it with thousands of methods. It was split across client code and several microservices. How did we monitor it?

We used a technique called activity-based monitoring. We first defined about 50 activity classes that represented important blocks of functionality in our system. Some examples of those activities included the following:

- RegisterForTTL: Tracks when a user registers to start receiving Time to Leave notifications

- UnregisterForTTL: Tracks when a user turns Time to Leave off

- StartTrackingEvent: Tracks when an upcoming event with a location begins to be monitored for travel time to it

- StopTrackingEvent: Tracks when an upcoming event with a location that was being monitored gets deleted or no longer has a location to route to

- ProcessMeetingWhenUserLocationChanged: Tracks when an upcoming event with a location has its location changed

- ChangeTransportationMode: Tracks when a user in the Outlook mobile changes their preferences for how they travel typically – users could choose to get routes by car or routes by transit if they used buses and trains to travel

– SendTTLNotification: Tracks when a Time to Leave notification is
triggered and sent to a user

Activities wrapped the execution of these logical units of functionality in the system.
An activity like "RegisterForTTL" would begin on the iPhone and would continue
through several microservices and then end back on the iPhone once the microservices
completed successfully. A TraceID was used to track the user's request from their
phone through the various backend systems and back to the phone – this TraceID was
just a GUID (a globally unique identifier that is a unique string for each user request
and looks something like "360C4783-E414-4CEC-A114-8513AA6A8CCE") that tracked
that particular action throughout the system. In C# the wrapping of an operation by an
activity looked like this:

```
Activities.RegisterForTTL.ExecuteAsync(async () =>
{
// Perform the operation
//
 return ...;
});
```

Each activity class had its own unique identifier, a long name like
ChangeTransportationMode, and a shorter abbreviated name (ChgTranMode). Activities
could also be nested. For example, the RegisterForTTL (RegForTTL) activity might have
nested in it several other related activities that happen as part of turning on the feature for
the first time – the StartTrackingEvent (StTrkEvnt) activity is one example of a potentially
nested activity that could be triggered if when the Time to Leave feature is turned on, an
upcoming meeting with a location is found that is close enough in time and far enough
away from the user's current location that it needs to start to be tracked to determine when
to alert the user. When Activities were nested, the system used the shorter abbreviated
names to create what was called an "activity vector" which represented the nesting of the
activity, so something like "RegForTTL > StTrkEvnt" shows that the StartTrackingEvent
activity (StTrkEvnt) was called within the RegisterForTTL activity (RegForTTL). These
activity vectors are emitted with each trace logged by the activity classes.

In addition, every REST entry point in our microservice API was also wrapped in an
activity. This allowed you to track execution from the point that a REST API was called on
the server through its successful (or unsuccessful) execution. There were on the order of
another 25 or so entry points to the system that were also wrapped with Activities.

Activities would then log events that occurred during the execution of the operation wrapped by each activity to a text-based logging system. The events that an activity would track were the following:

- ActivityStarted: When the activity began, a line of output would be emitted with the TraceID, the name of the microservice the activity ran within, the abbreviated activity name, the current "activity vector," and the event message text. For example, if the StartTrackingEvent occurred, there might be a line of logging emitted that looked like this:

  16f25a17d5d13aa0ba726a19a0d9801a4 | CortanaTimeToLeaveService | StTrkEvnt | RegForTTL>StTrkEvnt | ActivityStarted: Attempt=1

  The activity vector uses the abbreviated activity names to provide a compact logical call stack of the nesting of various activities contained within endpoint activities in the system. This helps to understand the flow of the system and allows powerful analysis as we will see later. As you can see in the message text, a retry system was built into some activities so they would automatically try to re-execute the code if a particular attempt failed for some reason and the number of retries was tracked in the message.

- ActivitySucceeded: When the activity completed successfully, a line of output would be emitted with the TraceID, the name of the microservice the activity ran within, the abbreviated activity name, the current "activity vector," and the event message text. For example, if the StartTrackingEvent finished successfully, there would be a line of logging emitted that looked like this:

  16f25a17d5d13aa0ba726a19a0d9801a4 | CortanaTimeToLeaveService | StTrkEvnt | RegForTTL>StTrkEvnt | ActivitySucceeded: duration=330, Attempt=1

- In this event, notice that the end to end duration of the activity is tracked from the time it started to the time it completed which in this case took 330 microseconds.

- ActivityFailed: If the activity failed for some reason, a line of output would be emitted with the TraceID, the name of the microservice the activity ran within, the abbreviated activity name, the current "activity vector," and the event message text with exception details and error message. For example, if the StartTrackingEvent threw an exception during execution, there would be a line of logging emitted that looked like this:

16f25a17d5d13aa0ba726a19a0d9801a4 | CortanaTimeToLeave Service | StTrkEvnt | RegForTTL>StTrkEvnt | ActivityFailed: duration=320, Attempt=1, Exception=Null Reference Exception in 'register.cs' line 239

In this event, notice that error details are reported that help developers understand what went wrong in the system and where.

- In addition to the three basics, "ActivityStarted," "ActivitySucceeded," and "ActivityFailed," custom messages could be emitted into the trace. These messages would be stored in the same basic format with Correlation ID, activity name, activity vector, and message; but in this case, the message would be the custom logging message defined by the developer.

Azure Data Explorer for Analyzing Traces

What really makes this system powerful is the addition of Azure Data Explorer to the mix. Azure Data Explorer is a log analytics cloud platform optimized for ad hoc big data queries of semi-structured text logs. All of the logs created by activity logging in the system were fed into Azure Data Explorer. With Azure Data Explorer, you can then do sophisticated ad hoc queries on the entire set of logs emitted by Activities. For example, you could look at just an arbitrary set of activity logs that have at least one level of nesting with a query like this:

```
8  traceevent
9  | where ActivityVector contains " > "
10 | take 5
11 | project Level, TraceId, ServiceId , ActivityName , ActivityVector , Message
12
```

Level	TraceId	ServiceId	ActivityName	ActivityVector	Message
4	16f25a17d5d14a0ba726a19a0d9801a4	CortanaTimeToLeaveService	UpdStiCrts	GetStiCrts > UpdStiCrts	ActivityStarted: Attempt=1
4	16f25a17d5d14a0ba726a19a0d9801a4	CortanaTimeToLeaveService	GetStiMnft	GetStiCrts > UpdStiCrts > GetStiMnft	ActivityStarted: Attempt=1
4	16f25a17d5d14a0ba726a19a0d9801a4	CortanaTimeToLeaveService	GetStiMnft	GetStiCrts > UpdStiCrts > GetStiMnft	ActivitySucceeded: Duration=322, Attempt=1
4	16f25a17d5d14a0ba726a19a0d9801a4	CortanaTimeToLeaveService	UpdStiCrts	GetStiCrts > UpdStiCrts	ActivitySucceeded: Duration=393, Attempt=1
4	36386981714149e28dafbc177ac46ead	CortanaTimeToLeaveService	GetStiMnft	GetStiCrts > UpdStiCrts > GetStiMnft	ActivitySucceeded: Duration=435, Attempt=1

You can also do more sophisticated aggregate queries with Azure Data Explorer. In this example, we use Azure Data Explorer to determine performance characteristics for a common activity vector that occurs in the system (GetStiCrts > UpdStiCrts). In this case, Azure Data Explorer uses the fact that we consistently emit "duration=" in the message of each ActivitySucceeded event and uses its "summarize" capability to calculate percentiles of performance for that particular sequence of activities across the entire system (GetStiCrts has to do with getting some needed Certificates in the system).

```
1  traceevent
2  | where EventInfo_Time > ago(7d)
3  | where ActivityVector == "GetStiCrts > UpdStiCrts"
4  | where Message startswith "ActivitySucceeded"
5  | extend duration = extract("ActivitySucceeded: Duration=(.*?),", 1, Message, typeof(int))
6  | summarize percentiles(duration, 50, 75, 90, 95, 99) by bin(EventInfo_Time, 5m)
```

EventInfo_Time	percentile_duration_50	percentile_duration_75	percentile_duration_90	percentile_duration_95	percentile_duration_99
2017-07-11 02:40:00.0000000	220	430	430	430	430
2017-07-11 02:45:00.0000000	432	494	494	494	494
2017-07-11 21:55:00.0000000	337	337	337	337	337
2017-07-11 22:00:00.0000000	205	301	301	301	301
2017-07-11 22:05:00.0000000	250	250	250	250	250
2017-07-11 22:25:00.0000000	378	378	378	378	378
2017-07-15 20:00:00.0000000	508	508	508	508	508
2017-07-15 20:10:00.0000000	312	312	312	312	312
2017-07-15 20:15:00.0000000	211	211	211	211	211
2017-07-15 20:35:00.0000000	295	295	295	295	295
2017-07-15 20:55:00.0000000	393	393	393	393	393

The combination of thoroughly covering your code with activity-based logging and using Azure Data Explorer to analyze that logging is a powerful combination that truly allows you to measure the progress of working software in multiple ways. If you haven't used Azure Data Explorer yet, prepare to be amazed – it is an extremely powerful tool, and as you begin to realize the power of both specific and aggregate queries on your monitoring data, it will completely reshape how you think about measuring your product. More information about Azure Data Explorer is available at aka.ms/kdocs.

Azure Data Explorer vs. a Database   You might wonder why you would use Azure Data Explorer over another database. Where Azure Data Explorer excels is in doing fast queries over terabytes of semi-structured text data. Similar products you could also consider include Splunk, Logstash, InfluxDB, or Elasticsearch.

What Monitoring Can Tell You

With activity-based monitoring in place and a system like Azure Data Explorer to query the resulting log data, you can now truly measure your product. Here are some areas you will want to dive into and begin to create Azure Data Explorer queries to analyze in your product.

Is the Working Software Really Working Software?

One immediate benefit of activity-based monitoring is you can now quantify how often your system is succeeding and how often your system is failing. This is subject of course to bugs or gaps in your activity monitoring, so make it a regular priority and part of your continuous integration process to ensure that gaps in your activity-based monitoring do not occur and as new functionality is added to the system, new Activities are defined to track that functionality. Assessing the health of the system is only an Azure Data Explorer query away. You can create an Azure Data Explorer query that tells you how many times an activity failed in the past day – something like this: In this example, our activity-based events are stored in an Azure Data Explorer table called "traceevent":

```
traceevent
| where EventInfo_Time > ago(1d)
| where Message startswith "ActivityFailed"
| count
```

You can put this in context with how many Activities succeeded with a similar query:

```
traceevent
| where EventInfo_Time > ago(1d)
| where Message startswith "ActivitySucceeded"
| count
```

You can dig into individual failures and what went wrong with a query like this:

```
traceevent
| where EventInfo_Time > ago(1d)
| where Message startswith "ActivityFailed"
| take 5
```

What Went Wrong?

You can dig into individual failures and what went wrong with a query like this:

```
traceevent
| where EventInfo_Time > ago(1d)
| where Message startswith "ActivityFailed"
| take 5
```

This will display five recent failures. You can grab the TraceID out of a failure and see the entire set of activities that happened during a session with an Azure Data Explorer query like this:

```
traceevent
| where TraceID == "paste the entire TraceID from a failure here"
| project TraceID, ServiceID, ActivityName, ActivityVector, Message
```

This will output a whole set of activity tracings with activity vectors so you can explore the exact sequence of events that happened before the failure occurred.

How Fast Is It?

As we saw earlier, you can examine a particular activity vector or entry point and see how long it is taking for most users with a query like this:

```
traceevent
| where EventInfo_Time > ago(7d)
| where ActivityVector == "RegForTTL > StTrkEvnt"
| where message startswith "ActivitySucceeded"
| extend duration = extract("ActivitySucceeded: Duration=(.*?),", 1,
  Message, typeof(int))
| summarize percentiles(duration, 50, 75, 90, 95, 99) by bin(EventInfo_
  Time, 5m)
```

You can look into a set of Activities that is taking a particularly long amount of time relative to other Activities with a query like this (which observes that the highest time in the previous query were in the 500 ms range):

```
traceevent
| where EventInfo_Time > ago(7d)
| where ActivityVector == "RegForTTL > StTrkEvnt"
| where message startswith "ActivitySucceeded"
| extend duration = extract("ActivitySucceeded: Duration=(.*?),", 1,
  Message, typeof(int))
| where duration > 500
```

If you suspect that the slower times are happening because of a particular issue – for example, maybe tracking an event is slower if that event is a recurring event – you can add to your logging a custom message something like "RecurringEvent," and then construct a new query to look for it. Note that this assumes that your activity-based logging implementation also emits duration when a custom event is logged:

```
traceevent
| where EventInfo_Time > ago(7d)
| where ActivityVector == "RegForTTL > StTrkEvnt"
| where message startswith "RecurringEvent"
| extend duration = extract("RecurringEvent: Duration=(.*?),", 1, Message,
  typeof(int))
| where duration > 500
```

Are the Business Goals Really Being Met?

As part of the design of your monitoring, you should assess what are the business metrics you will track to evaluate a successful project. One commonly used business metric is MAUs or monthly active users. Our definition of a MAU for Time To Leave was the number of users who were actively registered to receive notifications. This number proved to be somewhat meaningless for the business as it was quickly discovered through monitoring the system that there were a far larger number of people who turned the feature on than those who received a Time to Leave notification. Common reasons for a user receiving no notifications were they didn't enter the location for enough meetings on their calendar, they didn't have very many meetings on their calendar, or all meetings were close to where they were already located.

So it is a good business practice to define a second metric – a MEU (pronounced like a cat would say it). A MEU is a monthly engaged user. There are lots of ways to define this metric, but in the Time to Leave system, we consider an engaged user someone who had received at least one meeting alert from the system in the past month. The activity that tracked the sending of a notification was called "SendTTLNotification" or "SendTTLNot." A query to figure out how many TTL notifications were sent in the last month looked like this:

```
traceevent
| where EventInfo_Time > ago(30d)
| where ActivityVector == "SendTTLNotification"
| where message startswith "ActivitySucceeded"
| count
```

Of course, this query doesn't measure the monthly engaged users since one user could receive multiple notifications in a month. A more advanced query was needed to determine the number of unique users who received a notification.

Determining the number of unique users begins to get into the area of privacy – clearly the system knows the unique identity of the user, but we don't want that knowledge to be exposed to the logging as it represents a potential leak of information about the user to developers who shouldn't be able to make Azure Data Explorer queries and determine unique or private information about any individual user such as their email address, the name or location of a meeting, and so on. So to identify a user, a one-way hashing algorithm would be used to store a unique user ID calculated from uniquely identifying information about the user that the system has access to. Then this identity hash can be used to create an Azure Data Explorer query to determine the number of unique users in a month without leaking any personally identifying information into the logging system.

Aggregate and Anonymize   It is not necessary to store information about the user in perpetuity. For our system, we would purge the system of any logging data about a user after about 30 days. We would run regular Azure Data Explorer queries to store aggregate monthly data about key business metrics and other user metrics we wanted to track in the currently available logging data and store aggregate information longer.

Are the Customer's Needs Really Being Met?

Monitoring is also a useful way to determine if the customer's needs are really being met and, if not, why are they not being met. Through monitoring and logging, we were able to determine that one of the times many of our users experienced the Time to Leave feature was when they had a flight going out of the airport. This was surprising to us as we never did any special work to make this a priority. But we discovered that alerts to the airport happened frequently and they were useless to the user – airport alerts were being fired in a way that got the user to the airport 10 minutes before their flight was to leave.

As we investigated why this type of alert happened so frequently, we discovered an unexpected multiplier in a related system. Cortana had a system that would automatically add flight meetings to your calendar when it detected you received a flight confirmation email in your inbox. Cortana created the flight meetings with the address of the airport. So now we understood why there were so many of these meetings on people's calendars that were causing Time to Leave notifications to be raised.

We then worked with the team that built the inference system that created these meetings automatically so that we could recognize these meetings as flights. We then modified our system to fire alerts for a flight such that the user would arrive at the airport 2 hours before the flight departed from the airport.

How Are the Data and Models Being Used?

It is also useful to consider ways to monitor the data that your system produces and presents to customers as well as the models that run in production. There are important things to be learned with effective monitoring.

One surprisingly useful piece of data that we collected in Bing's local data team was we recorded how often a particular business was actually displayed to a user. We would record each time a local data entity was shown on the search results page, each time it was displayed as a top of page answer, each time it was displayed in a map, each time the detailed description of a local data entity was shown, and so on. This gave us a very good idea of which businesses were actually being displayed and surfaced by our search engine. Close monitoring of how often businesses were being displayed helped us to track the types of businesses users cared the most about. We were able to determine which kinds of businesses were getting searched for the most and invest in those categories. For example, we had an internal list of categories that had the most number of views, and we made decisions of where to focus our efforts based on the categories that were being searched for and displayed most often. We would also track categories that were rising the most rapidly month over month. As an example, this led us to make the food trucks category a focus during one particular year when that category was rising rapidly.

Monitoring which businesses were being displayed and how often businesses were being displayed also helped us detect issues in our site. We could track month over month which entities in our system were getting a lot of views, and sudden changes would be flagged for review. By monitoring display counts for entities, we detected that one of our frequently viewed entities – a casino in Las Vegas – suddenly went from tens of thousands of views per month to no views. Upon investigating, we discovered that our conflation system had overmatched the casino to a restaurant with a similar name and so the casino had completely disappeared from our system. We also could detect abuse of the system through tracking suspicious entities that began from zero views and started to amass thousands of views per month. These were sometimes fraudulent businesses that had been submitted to us and slipped through the verification process.

Often, a machine learning model will be running in production and must be carefully monitored for changes to its performance. The canonical example are models that are designed to detect fraud and thereby have built into the problem space an adversarial relationship with an external bad guy. The bad guy wants to figure out how to get around a fraud detection model and deceive the system. It is important to monitor over time the number of fraudulent transactions a model detects, and the number of transactions it determines includes no fraud. If the ratio of good transactions to fraudulent transactions changes significantly over several months, it is unlikely that it is a result of evil reducing in the world and the fraud going away. It is much more likely that the bad actor has determined new ways to defeat the system. Models subject to adversaries need to be monitored and frequently retrained and enhanced.

Conclusion

In Chapter 7, "Monitoring", we discussed how activity-based monitoring can be used to truly measure your working software. We would rewrite this agile principle as "Monitoring of working software and data provides the primary measure of progress." We discussed how an activity-based monitoring system works and the kinds of things that it logs and tracks. We talked about Azure Data Explorer which allows you to do sophisticated ad hoc queries of large amounts of logging data. We examined the various kinds of things that good monitoring can tell you: about whether your software is really working as expected, about why and when the software is malfunctioning, about what performance users are actually experiencing from your system, about whether the business goals are being met, about whether your customers' needs are being met, and about how data and machine-learned models can be monitored in production.

In Chapter 8, "Sustainable Development", we will talk about how to determine if you are working too fast or too slow and how to adjust the pace down and up. We'll also talk about goal setting and keeping team engagement high.

Sustainable Development

*Agile processes promote **sustainable development**. The sponsors, developers, and users should be able to maintain a constant pace indefinitely.*

—agilemanifesto.org/principles

VentureBeat.Com reporter Alex Banayan describes Qi Lu, executive vice president at Microsoft of Bing:

> *Growing up in a village outside of Shanghai with no running water or electricity, Qi Lu (pronounced: chee loo) had no idea that one day he would have a corner office at one of the world's biggest technology companies. As the President of Online Services at Microsoft, Lu has made a drastic journey to the top thanks to what his colleagues call "Qi Time."*

> *"During college, the amount of time I spent sleeping really started to bother me," Lu explained to me. "There are so many books I can read and so many things to learn. It feels like, for humans, 20% of our time is wasted [during sleep] in the sense that you're not putting that time towards a purpose that you care about."*

> *Although he admits it wasn't easy, Lu has engineered his body to function on four hours of sleep a night thanks to an unusual regimen that ranges from timed cold showers to daily three-mile runs.*

> *Driven by an unusual hunger to do more, Lu's sleeping schedule has added an extra day's worth of work time per week, which aggregates to nearly two months of productivity latched on to every calendar year. And he did it while still in college.*

As part of Bing Local, we worked in Qi's organization and witnessed his amazing ability to work long hours. Colleagues would see Qi running laps through the building early in the morning or on weekends and holidays. Ultimately, Qi's amazing ability

© Eric Carter, Matthew Hurst 2019
E. Carter and M. Hurst, *Agile Machine Learning*, https://doi.org/10.1007/978-1-4842-5107-2_8

to work certainly drove a certain amount of urgency and longer days into the Bing organization. There was no organizational expectation that anyone worked as long hours as Qi worked, but his ability to work long hours certainly drove a culture of people aspiring to work more than a basic 9 to 5 schedule.

But more than Qi's long hours, the rallying cry of trying to "beat Google" was probably a greater motivating factor that naturally drove people, especially the competitive ones, to work longer hours. A love for the technology and fascination for the problems being worked on also combined to drive just the right amount of engagement and urgency.

On other teams we've worked on, we've witnessed the opposite problem – a lack of urgency and people appearing to work shorter days. It is difficult to measure the intensity and amount of work being done just by how long people are in the office in a world of telecommuting and VPNs where anyone can be at work at any given time of day. But on teams where the urgency wasn't high, common issues that drove this opposite end of sustainable development were lack of vision, poor communication about schedules and expectation of what needed to be completed when, and programming environments and developer experience that was frustrating and unrewarding.

Are We on the Right Sustainable Pace?

So how do you determine whether the pace the team is on is too fast and unsustainable or too slow and sustainable? Typically, it is much easier determining the team is on an unsustainable pace. Common signs include not just continually long days and long weeks but short tempers, illness on the team, poor decision making, decreased morale, and fatalism intermixed with black humor.

Determining that the team is on a pace that is too slow is trickier. Common signs include not just continually short days and short weeks but short tempers, illness on the team, poor decision making, decreased morale, and fatalism intermixed with black humor.

Yes, that is not a misprint – the signs for being on a sustainable pace and an unsustainable pace are often the same. A better way to measure whether you are on the right sustainable pace is determining whether people are engaged and enthusiastic about the work. Engagement and enthusiasm will naturally generate an amount of work.

Microsoft has an annual poll in which they measure among other things something that they call an "Engagement Index." The Engagement Index is a combination of several components, but enthusiasm is gauged by questions like these:

- Do people go beyond their day-to-day work responsibilities to help the team succeed?

- Are you excited to come to work most days?

- Do you spend most of your time doing work you truly enjoy?

When teams are doing work they are excited about and enjoy, the pace they maintain will be sustainable.

Adjusting the Pace Down

Unsustainable paces are ones that are forced by edicts or demands of the business that can only be met through long hours. Clearly, you can't always avoid these types of things coming up – but you can work to try to eliminate edicts and unreasonable demands. Typically this is done through communicating better to the business what is reasonably challenging to complete in a period of time vs. what is impossible. Teams that truly understand the incremental development of software will work hard to establish meaningful regular goals that are both challenging and also achievable.

In machine learning projects, one challenging thing that must be understood by the entire team including the business is the rate at which improvements can be made to models. Typically, early on in a development process, models improve quickly to a particular level. But once models achieve a high level of performance, future movement is more difficult. It will often cost twice as much to move a model from 90% accuracy to 95% accuracy as it did to move the model from 70% accuracy to 90% accuracy.

One way to help the business understand this is to regularly review with them the improvements to measurements being made of models being shipped and to plot those improvements on a time line. Over time, with repeated exposure to the rates of progress made in particular domains, the team and business will gain confidence about how much improvement to expect in a given period of time. Possibly some rules of thumb will emerge.

In Bing, in areas where we were behind Google in terms of precision and recall of our models, we developed a rule of thumb over time that we could typically halve the gap with Google during a 6-month period. If a goal for improving the models fell behind this level, we pushed to increase it. Unfortunately, this law also applied in reverse – in areas where we were ahead of Google, they typically could halve the gap in a 6-month period of time as well. So in areas where we were ahead, it was important to continue to make progress as well.

Another rule of thumb was that the closer you would get to 100% precision and 100% recall (obviously not possible), the more skeptical you would be about making multipoint climbs within a period of time. It was not impossible to go from 97% precision and 97% recall to 98% precision and 98% recall, but everyone had a healthy respect of how challenging it would be to gain a point like that in the face of our noisy data sources.

Demos Can Be Dangerous   A pitfall teams often fall into is showing the flashy but ultimately unattainable demo to the business that indicates that something that may take months to achieve is achievable in weeks. Be very careful when demoing features that have massive amounts of work to be able to apply generally to all possible inputs or that will require a long amount of time to be performant, scalable, compliant, or architecturally sound enough to turn into real product. We have all too often been in a scenario where a half-baked demo communicated to the business a feature that seemed to be eminent only to find it took months or even years to deliver on the promise shown in the demo.

Adjusting the Pace Up

Everything we said in the previous section also applies to adjusting the pace up. Metric goals can also be not aggressive enough. In an area where a goal was significantly below a rule of thumb, the team pushed themselves to increase the goal. If it became clear that a goal was going to be met in a shorter period of time than was originally expected, the goal would be revised upward for the 6-month period.

It is also important to address the root cause of lack of engagement that is at the heart of a low pace. Dig in more to why people don't go beyond their day-to-day work responsibilities to help the team succeed. Common reasons may be that people don't feel empowered to make decisions on their own or they feel left out of team decisions and design discussions. People on the team may not feel they were included in the goal-setting process. Having a mixture of top-down goals and bottoms-up goals can help the team to feel more engaged.

If a significant portion of the team is reporting they aren't excited to come to work most days, the team should retrospect on what is causing this lack of excitement. Is it the developer experience? Allocate some development time to focus on the issues in the developer experience and improve tools and processes. Are people not excited by the vision? Try to clarify and help people see what the team is trying to achieve and why it could make a big difference for the world. Sometimes you may find that the kind of work people are excited about just won't be available in your area – this is an opportunity to encourage people to find positions where they can do work they are excited about and recruit new people who have a natural affinity for and excitement for your problem domain.

Occasional edicts and harder demands are not all bad either. Sometimes a hardship date or an edict that something has to be done or the business is at risk can help a team to coalesce and move toward an "we're all in this together" mentality. As an example, Microsoft and the entire computer industry faced a hard deadline on May 25 of 2018 to make sure all services were GDPR compliant. As that deadline approached and teams realized how much there was to do to meet the GDPR guidelines, we witnessed teams around Microsoft rally and rise to the challenge. During that period, engagement increased despite the longer hours because people could truly see they were involved in something worthwhile that they could be proud of – ensuring privacy for the individual against big tech.

The Importance of Changes of Pace

Even a high-performing team on a sustainable pace needs changes of pace on a regular basis. We recommend two practices to break up the pace and change things up on a frequent basis.

The first practice to break up the pace is what is known in agile as "slack weeks." Slack weeks are an opportunity to take a week, maybe 2, and give people a chance to work toward the secondary goals that seem to continually get pushed aside to hit milestones and schedules. Common activities during slack weeks can include spending more time investing in improving dev processes and tooling, reducing technical debt in the product, refactoring or rewriting problematic portions of the product that are slowing people down, and training and app building on new technologies the team is thinking about using. We typically have slack weeks around holiday times when teams aren't at full capacity anyways (July and December) and then find time a couple more times during the year to insert slack weeks into the schedule.

When the team knows a slack week is coming on a regular basis, they can queue up work items in their backlog that are targeted for that slack week – work that just doesn't seem to get prioritized high enough to be scheduled otherwise but the team still agrees is worth doing.

During a slack week, we usually will stop much of the scrum processes described in Chapter 4: Aligning with the Business. We will still hold a daily standup, but the idea is to give a change of pace and give people a break from "business as usual." Just make sure you do return to business as usual when you start your next sprint.

A second practice to change pace is to have regular hackathons. These are typically shorter-lived changes of pace – maybe two to three days repeated three to four times a year. These break up the pace in a different type of way – during a hackathon, the team is encouraged to work on some different problem area or idea but in a much more intense way than they would normally work. So whereas a slack week is a time to slow down, a hackathon is where the emphasis is on speeding up and making a lot of progress in a short amount of time on areas where architecture spikes need to be made to understand whether the team should invest further in that direction. During a hackathon, teams encourage longer hours through having more aggressive goals in terms of what gets done during the hackathon but with the additional expectation that anything that is done during the hackathon is exploratory, lower quality, and "Hackey" and will be thrown away after the hackathon (although the hackathon might result in a new product idea or direction that would then be reimplemented in a proper higher-quality way). hackathons are great opportunities to work fast and dirty for a change and maybe in the process discover significant new product directions that can be pursued later in a more quality way.

For hackathons, the team will basically think as part of their regular sprint planning about some area or feature that they are interested in exploring but they think is high risk. But they want to spend some intense time investigating that new direction. They will then schedule a 2- to 3-day period to completely context switch into that new direction and see how much they can prove out in a quick and dirty way during that period of time.

A team's half-year schedule might look like this. This schedule alternates regular paced two week sprints with slower paced slack weeks and faster paced hackathon weeks:

- Slack week

- Four 2-week sprints

- Half-week hackathon

- Four 2-week sprints

- Slack week

- Four 2-week sprints

- Half-week hackathon

Live Site and Sustainable Pace

One of the things that can most easily move a team from a sustainable pace to a non-sustainable one and can cause burn out the quickest are live site issues. Live site issues can disturb sleep and schedules as teams frantically try to figure out why the product isn't working in production.

If you are fortunate enough to work on a product that doesn't have a live site component, then you may not relate to this topic. But in a world of 24 × 7 services aiming for four nines of availability, live site issues can cause teams to melt down.

We will briefly share some best practices used by Bing around managing live site and making sure that it doesn't lead to unsustainable pace on your team.

In some teams, there may be a smaller operations team that handles all live site issues for a service. But in Bing and many other online engineering organizations, allowing all engineers to witness and troubleshoot what can go wrong on the live site is an important learning opportunity that leads to better code and better architectures long term. There is nothing like having to get up at 2 AM in the morning to fix an issue caused by buggy code you checked in or an architecture that you thought was going to scale but now doesn't work properly in production to motivate you to figure out a better way to do it next time.

So to that end, engineers typically participate in a rotation program where they are "on call" for a particular week and are the first responder to any live site issue during that week. Typically that "on call" week is prefaced by an additional week where they are "on backup" for the person who is on call. This allows the on call developer to be aware of what happened the previous week so when they are fully in charge of issues for the subsequent week, there is continuity in understanding what is going on in the web site since issues from the previous week can easily recur or cause some new side effect the subsequent week.

On call developers should have lots of available documentation on how to troubleshoot and detect and resolve issues in various parts of the system. Diagnostic tools and playbooks to use them were prepared so even new developers on the team could follow a set of basic steps to determine what was going wrong and, if not the root cause, at least a quick fix to bring the site back to its previously functioning status.

In Bing, we also kept a close eye on three metrics. The first was time to detect. This measured how long it took us from something going wrong on the live site to an automated system or customer detecting it. The second metric we measured was time to engage. This measured how long it took us from knowing something was going wrong to having an engineer engaged start to figure out what was wrong. The final metric we measured was time to mitigate. This measured how long it took us to apply enough of a fix – sometimes a quick fix or a rollback – so that the issue was resolved.

All three metrics, time to detect, time to engage, and time to mitigate, were monitored and continually discussed to figure out new ways to shorten each time. Goals were set to reduce and improve these three numbers each quarter. To improve time to detect, we often deployed additional monitoring or found new ways to detect a potential failure before it ever got deployed to the live site. In time to engage, we typically would improve processes to ensure the right engineer answered the phone as quickly as possible. In time to mitigate, we devised new ways of quickly rolling back or switching off new features in production that were causing problems.

On a weekly basis, it is important to postmortem all the live site issues that occurred during the previous week and try to figure out specific countermeasures to prevent them from happening again. These measures may include anything from architecture changes to process changes to better monitoring to additional documentation for on call engineers.

In the postmortem process, we would prioritize live site events to discuss by severity based on customer impact and high lengths of time to detect/engage/mitigate. Impact on the customer was investigated further to understand how many users were impacted by the issue and what the actual impact was on the user experience on the web site. Root causes for live site events would be determined by the engineering teams. Then items to repair would be identified and followed up on until they were fixed.

If live site training, involvement by engineers, documentation, rotations, postmortems, and follow-up on repair items are rigorously followed, the unsustainable pace of continual live site issues can be considerably reduced. It provides a virtuous cycle that ensures that everyone on the team thinks closely about the code and architectures they use to minimize those late-night phone calls.

Sustainable Pace and Multiple Development Geographies

A final area of challenge in sustainable pace is working with multiple development geographies. This can be sustainable if set up right, but can be a sustainable nightmare if set up in the wrong ways.

In Bing, we worked with development teams in India and China which were on substantially different time zones than the US offices in Seattle. Working across these time zones can become painful for teams and cause both unsustainable fast paces and unhealthy slow paces.

The unsustainable fast paces typically occur when there is a concept of the "main team" and the "remote team," and everything the remote team does has to be signed off on by the main team. This leads to the remote team sitting idle waiting for approval of work they want to do and the main team feeling overwhelmed and on an unsustainable pace because of all the remote hand holding they have to continually do in the evening or early morning hours of their workdays.

The solution to this, of course, is to not have the concept of a main team and a remote team. All teams should be empowered to fully own the area that they are working on and be able to make all of their own choices during their own business hours. Points of integration can be specified via interface design and building in natural points of transfer of information and concerns between system areas owned by the different teams.

Remote Teams Can Make for More Sustainable Pace   Teams in multiple geographies can also help to improve the sustainability of live site engineering. A team in one geography can man the live site during their daytime hours, while a team in another geography and time zone can get some much-needed rest. Then the teams can exchange the favor, taking over live site responsibilities when the sun comes up in their geography.

Conclusion

In Chapter 8, "Sustainable Development", we talked about how to determine if you are working too fast or too slow and how to adjust the pace down and up. We would rewrite this agile principle as "Agile processes promote **sustainable development**. The sponsors, developers, and users should be able to maintain a constant pace indefinitely, but varying the pace up and down occasionally can produce an even better result." We talked about how to set goals and ensure that goals are both sufficiently ambitious and also achievable. We talked about the importance of engagement on a team to get the pace to the right level. We also talked about the importance of having some changes of pace built into development schedules with slack weeks to slow down the pace and hackathons to speed up the pace. We also talked about special issues around sustainable pace in managing live site issues and in working with multiple geographies.

In Chapter 9, "Technical Excellence", we will discuss how an investment in excellence in software engineering can optimize developer productivity while maximizing the value of data.

CHAPTER 9

Technical Excellence

*Continuous attention to technical excellence and **good design** enhances agility.*

— *agilemanifesto.org/principles*

When we think of agility, we think of efficiency of movement and an ability to change direction in the face of a complex and changing terrain all in the pursuit of a specific goal. Usain Bolt running 100 meters in 9.58 seconds is impressive, but it isn't an expression of agility. Lionel Messi, on the other hand, weaving in and out of a lattice of defenders while keeping the soccer ball constantly under his influence is more like it.[1] Agility requires quick movement for sure, but it also requires the behaviors that allow us to do this without distraction (Messi doesn't walk onto the pitch with his bootlaces undone); it requires the mobility, or dexterity, to manipulate the environment or change course while maintaining stability; it requires the awareness and presence of mind to read the shifting terrain and environment in real time to anticipate and adjust without missing a step.

Many developers confuse the desire for quick progress with the fundamentals of agility. Agility is compounded efficiency. It's the effects of low-level interactions on a network of processes and people. Much of what an engineer does should yield a multiplier in return. You write a test, and you save every developer who uses your code hours of time debugging complex errors in downstream systems; you automate deployment, and you reduce the time taken to ship your update by an order of magnitude; you enable continuous integration and save everyone who clones your code countless headaches in debugging the build process; you invest in design, and developers will use your system for applications you didn't imagine.

---

[1]Messi is also good at the tenth principle – maximizing the amount of work not done – which is why he often *appears* to be wandering around aimlessly in midfield.

© Eric Carter, Matthew Hurst 2019
E. Carter and M. Hurst, *Agile Machine Learning*, https://doi.org/10.1007/978-1-4842-5107-2_9

It can be difficult to take the time and effort to invest in efficiency when it is not directly on the path to shipping code. Why build a labeling tool when we can copy and paste data in a couple of hours? Why write tests when it obviously works and I could get on to the next task? We recognize that there is something of a leap of faith here – if you haven't both been bitten by the downside of cutting corners and experienced the wins of efficiencies built into your culture, it can be hard to make that leap.

In this chapter, we will examine a number of scenarios in which attention to technical excellence and good design will help you build a culture of efficient, sustained development and delivery of data projects. We will do this by first persuading you that investment in quality is good for agility, and then we will discuss what investment in quality means for data projects. Before we do, a word on *continuous attention*. This term is really a signal to indicate that it takes work, focus, and not a little cajoling to get a team to improve on these behaviors. As a leader, you have to be looking for ways to demonstrate to your team that these really are good things to do. As a developer, you have to keep an open mind and be willing to invest in something that will cost you up front but will pay you and your team a dividend down the line.

Software Engineering Practices for Agility

One way to think about agile teams is via the notion of *developer productivity*. Every tool, behavior, and process that you put in place should be primarily concerned with making everyone productive. Of course, productivity is not sufficient – it needs to be focused and guided, but we find the mantra "optimize for developer productivity" to be pithy and effective as a general guide. If your team can't fundamentally get things done in an efficient manner then, well, you're never going to ship. If you're managing a team, you want to be continuously figuring out if you are enabling and investing in your team's ability to get stuff done.

When the agile manifesto principles talk about the role of technical excellence, it is addressing a core truth of the engineering discipline: all actions and decisions have a network effect through the team and product. Good actions are those that have positive multiplier effects as they cascade through the network. A good API design makes it easier and safer for developers to build on top of; good planning makes it easier to hit goals.

Let's think about these network effects in terms of unit testing – a great example of technical excellence for general software engineering. Unit testing not only contributes to an individual's productivity but also the entire team's as well as that of partners and collaborators. When writing code, starting with a test (i.e., practicing test-driven development) enables the developer to get clarity on the exact nature of the function to be computed. There is a moment where any lack of clarity can be circled back to others on the team or the customer. The developer can think first about what success looks like. Writing the code then becomes an exercise in satisfying the test. This modularizes the work – the tests become the manifestation of the requirements. The code is then submitted for some form of review. A reviewer can look at the test and get an idea of what the function should be doing (they will also review the test to ensure that it is meaningful and useful). Once the code is checked in, it may be used, extended, or modified by another developer. Because tests are in place, that developer can be sure that others using the code are protected (if their tests break, then something has gone wrong, and the semantics of the function are no longer intact). The investment in testing pays off for the developer, the code reviewer, the immediate team, and partner teams.

The fun doesn't stop there – if testing is embraced, then it means that code is *designed* for testing. If code is designed for testing, then units of computation are kept small and modular – making them understandable to reviewers and maintainers; testing can enhance the abstractions and encapsulation of object oriented-programming; when it comes to refactoring, a built-out set of tests will help guarantee that the code may have changed but the functionality is preserved.

You will sometimes hear people say "Yes, but you can still write bad tests" or "Tests give a false sense of security." That is like saying that just because a rock climber has a safety harness on, they could still fall. It turns out that rock climbers are well motivated to ensure that their safety equipment works – and so should software engineers be motivated to ensure that their code works – through testing. Writing good tests, like writing good code, is a skill. Those who embrace it will develop that skill and help their peers improve also.

Other areas where we can see the positive network effects through technical excellence include the following:

- Continuous integration: When a change is submitted for review, it should also trigger a build. This build should include running all tests in the repository. This allows the team to get immediate feedback about obvious defects at no cost. Continuous integration is offered as a standard feature on Azure DevOps.

- Push button deployment: When your code is deployed to a number of targets (a production service, testing tools, applications), it can make a release a tedious business if each step is manual. Whenever teams are faced with mundane tasks to perform, they will find other things to do. The result is inconsistent and error-prone releases – your test tool has a different version of the code to that currently running in production. Providing your team with a button that automates the whole thing (including validation tests) is a big win.

- Code hygiene: Teams that practice good code hygiene avoid copying code, invest in packaging, avoid checking in binaries, avoid checking in auto-generated classes, follow coding style guidelines consistently, write comments, and so on. Many of these benefits come with an upfront cost – but that down payment is tiny once you start benefitting from the results.

These techniques are discussed in more detail in Chapter 3: Continuous Delivery. In summary, technical excellence in general software development helps maximize developer productivity by removing friction points.

The second clause of principle 9 addresses good design. Technical excellence is, perhaps, easy to prescribe. But what is good design? Plenty has been written on this topic, often distilled as distinct principles or groups of principles: the SOLID principles (attributed to Robert C. Martin), the KISS principle (attributed to the US Navy), and the Einstein principle ("everything should be made as simple as possible but no simpler"). These are often complemented by process guidance principles such as building a Zero Feature Release (ZFR) or establishing a Minimal Viable Product (MVP). In our experience, establishing the team's attitude to certain engineering practices can really help.

The first is the attitude to programming languages. On the one hand, you can fix the language so that all developers share a common point of reference. This makes code reviews, testing, and collaboration easier. At the other extreme is complete flexibility around languages. This allows for better selection of the right tool for the job, but has the downside of less concentration of expertise. In Bing's local data team, we generally decided to stick to object-oriented programming, specifically using C#, while leveraging other systems for scripting and some non-production code. Object-oriented programming offers a compelling framework that is flexible enough to model data and processes but, with some discipline, can be constrained enough to help guide design through well-established patterns. The core principles of object-oriented design are abstraction, encapsulation, inheritance, and polymorphism. Every design and code review can be inspected for these principles, and the team can share them as a framework for discussion. There are other options to be considered – functional programming is enjoying something of an upswing, and less structured coding styles such as those enabled by Python are in fashion.

The second is the team's attitude toward design. In our web mining team, we developed the expression *a code review is not a design review unless it is*. What this means is if you don't run a design review for your feature, then you are at the mercy of the reviewers who may hold up the code review not on code quality issues, but on design quality issues.

The third is a commitment to a specific operational paradigm. This is similar to the team's attitude to programming languages but involves the choice around distributed compute and storage and can very much depend on the system context in which you are working. For example, if your company has built out a large proprietary infrastructure around MapReduce, then you are probably going to be designing to that paradigm. On the other hand, if you find yourself in a service-oriented culture, then your solution will be quite different. While one approach might be better than the other, there are two things to keep in mind. Firstly, ensure that your core computational asset is built in libraries that are not dependent on the operational paradigm; secondly, isolate the optimizations for the operational paradigm from the optimizations for the core functionality. This will allow you to be flexible when the inevitable switch happens without incurring any significant upfront cost.

Be Open but Commit   There are an almost infinite number of choices regarding coding and platforms, and good developers and managers need to be aware of the space as it evolves. However, decisions and commitment are required if the team is to have a hope of delivering. It is not acceptable to be closed to those options – technical leads and managers, in fact, should be continuously scouting and keeping up to date; but it is also foolhardy to constantly change to something that might be better. The start of a project is a good time to review options and experiment, and architectural modularity and decision deferment are good practices to help the team remain open.

Technical Excellence for Data Projects

Considering developer productivity and the network effect of good (and bad) contributions gave us an idea of the role of technical excellence for software engineering in general. How does this translate to the data aspects of data projects? Our discussion of network effects explains why technical excellence is important. The parallel to "optimize for developer productivity" on the data project side is to "maximize the value of data assets." Let's break this down.

As we progress through a data project, we generate many data assets. These include schema, samples, labeled data, models, analytics (e.g., the performance of a model on a test set), comparisons, and so on. The value of the data can be thought of in two ways: the immediate value and the long-term value. The immediate value is the reason it was created – it solves a problem, provides knowledge, or somehow moves the project forward. For example, a labeled data set provides value by allowing us to train a model which can then, if acceptable, be shipped to production. The long-term value, however, is the value we get out of the data later. This may be planned or unplanned. Planned long-term value may be, for example, the continuous use of a ground truth data set for a metric. Unplanned long-term value describes the value we get that is unlooked for. For example, we may benefit from an old model when running comparisons against new methods using a new test set (one not available when the old model was created). We may want to apply a new method to cleanse labels to an older label set to see if it improves on the models originally generated. Value can also be found in the systems and tools that we use to process and inspect our data, for example, sampling scripts, custom labeling and data browsing tools, and data cleanup workflows.

Perhaps surprisingly, it is common for data projects to underinvest in ensuring the planned for long-term value of data, let alone invest in the potential of unplanned for value.

You Are What You Measure

If you are working on a reasonably interesting problem, you probably have a pipeline or other form of process workflow with several stages. Data can be harvested at multiple points in the pipeline. When it comes to metrics, it is important that you are clear on what you are measuring. This is especially true if you are reporting official metrics that a partner or customer team will rely on. It is essential that you are clear on what your *data product* is and that you measure just that.

A common model for offline data pipelines, and one that we adopted for Bing's local search product, is to deliver an XML product. XML is used due to its ubiquity (any platform can consume it), its standards (there is little ambiguity regarding what is correct XML), and its reasonable human readability.[2] This XML format was what we sampled from and what was used in our measurement system. Many consumers of the system required the data in a different format – especially those that had clear requirements for space (XML is verbose) and computational performance. For those customers, either they wrote their own transformations of the data or we provided them. This final serialization is something that can be rigorously tested to ensure that it doesn't in any way undermine the statements that you make based on your production metrics. This means that the measurement system pulls data from a well-defined bottleneck point in your system from which all views of your data emanate.

With this setup, you can also build validation processes that take the data product output for any input and validate the other versions of your data that you may be providing for your users. For example, you can take the fields of an XML document and see if they are preserved in your space-efficient, binary serialization.

There are many reasons why you may be tempted to take some other view of your data. It may be simpler due to your architecture to harvest the data prior to it being written at the bottleneck stage; it may be that the form in which the user consumes the data doesn't work well with an existing measurement pipeline. It may be because the output format has changed and you built your measurement system to consume the legacy format, and it lacks the flexibility to retarget to the new format. Whatever the reason, you should resist this temptation at all costs. If you do anything other than measure a well-defined product, you run the risk of a transformation which introduces some change that renders your metrics incorrect. Figure 9-1 illustrates these different measurement scenarios.

---

[2]JSON is another format with these qualities. Technically, XML is a markup language – something that is interleaved with object content – so JSON would have been a better option to capture a data structure.

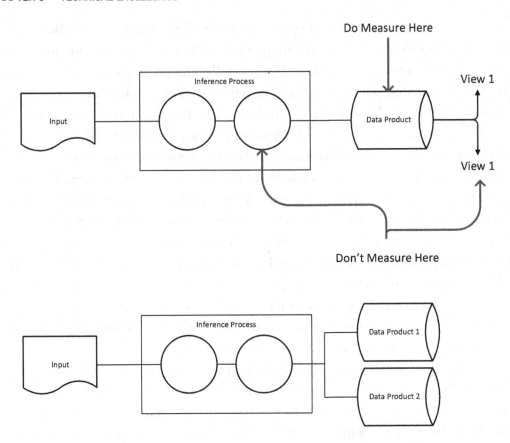

Figure 9-1. *Measure this, not that – In the upper diagram, the system produces a singular data product which is a reliable place to measure quality. Quality should not be measured at a stage prior to the data product nor from different, client-specific views of the data product. In the lower diagram, two different data products are produced with no singular artifact, and so no reliable measurements can be made.*

Avoid Unintentional Transformations[3]   As your data passes through your system, it undergoes one or more transformations. As the term suggests, intentional transformations are exactly those which you want the system to

---

[3]Also good advice for werewolves.

execute. They are adding the value. Unintentional transformations are those that have a material effect on the data which is either unknown or about which your assumptions don't hold. These can happen in any number of places, but two of the most common are data serialization and data labeling. When we serialize data, we have to transform it from an in-memory format to a disc (or stream) format. If we don't pay attention, then errors can creep in here – encoding errors (for textual data), data structure loss (e.g., turning a tree structure into a sequential form), reducing color depth (for images), and so on.

When we are labeling data, we need to show it to the user somehow. This may involve some form of serialization, but transformations may also occur in the labeling tool and even the display device that being used by the labeler. Labeling web data is notorious for this problem. Web pages are rendered for the user when the browser downloads the bytes from a URL, applies the associated CSS styling, and runs any JavaScript that is present on the page. All of this can transform the data considerably, and so when we save web data, we must go to great lengths to ensure that it is persisted in a format that allows for it to be viewed accurately later on.

Another way to think about this issue is to understand that your data product should primarily be designed for you – not your customer. This may sound counterintuitive given the focus we all have on the customer. What the customer consumes is a specific view, a serialization, of the data product. Your data product needs to take into consideration the development team's needs – transforming it to something for the customer should be a last-mile, simple, and well-tested step. Excellence in measurement, then, means measuring what you produce and ensuring that downstream, last-mile serializations do not transform your data.

We've talked about the details of the output of your system and how it relates to the measurement process. Let's talk now about the input in the context of measurement. It seems intuitive to require that the input to your labelling system is identical to the input to your process. This brings a certain coherence to all the components. However, when humans are judging or labeling the data, there are legitimate reasons why the view of the data that they see may be different from that which is the input to your system. Imagine an inference pipeline tasked with classifying images. The classifier has to determine if the image contains a mountain lion. Because this system is going to be deployed with many low-cost digital cameras, the budget dictates that the cameras must all be black and white. While developing the judgment tools to create labeled data, you discover that the judges have a hard time classifying the images due to the lack of color. To improve the process for creating labeled data, it makes sense to show the judges the richer color images.

Developing Models While Building Metrics

Congratulations! The project that you pitched has been funded, and you are about to start work with your small team to deliver it. You have 4 weeks to go from zero to the first live deployment. You have no metrics. You cost out the time taken to define the metrics, implement the tools, train the judges, and integrate the dev set into your inner loop and discover that it will take around 3 weeks to do so. This would leave you no time to work on your inference components in the ideal metric-driven manner. What do you do?

Even for teams that have some form of metrics framework in place, a new metric can take time to develop. The concept being measured will take multiple iterations between writing guidelines for judges, training them, reviewing their output, and adjusting the guidelines to accommodate emergent ambiguity or confusion. While the metric-driven approach should always be preferred, you will find that often your team will have to get started without them.

The trick to navigating this stage of the project is to rely on the fact that major problems (i.e., challenges in your inference problem that are worth addressing) should be observable in a small sample of data. This means that devs can spend an hour or 2 manually labeling data to find areas of improvement that have a high probability of impacting the final metric. If 20% of your system's output suffers from a specific issue, you should be confident in observing this through the inspection of a few 100 examples.

Writing Tests for Inference Systems

When developing a machine-learned model to deliver some form of inference while working with the same evaluation data, we will generate any number of candidate models. As these models progress toward an acceptable performance in terms of the precision and recall we measure, there will be variations in their output. Many of the examples in our evaluation set will produce the same results (these are the easier cases captured by the model). However, net improvements will also result in some marginal regressions where good results will flip to being incorrect. The same is true across releases of the system – we will generally be looking for net improvements and make calls about regressions of previously correct results. With procedural systems, unless the semantics of the functions being implemented change, we would not expect to see tests failing as we make progress unless a real bug has appeared. However, if we were to test inference systems in the same way (given an input, expect to always get the same output), we would be dealing with tests that flip-flop between passing and failing. Let's walk through a typical inference pipeline and take a look at how we might go about writing tests.

An obvious place to start is at the feature generation phase.[4] Computing features for an input is a textbook scenario for writing tests. Each feature being computed should be tested, including tests for cases where the feature is not present in the input. There is nothing worse than discovering a bug in feature computation for a deployed system only to find that fixing the feature – which will only impact the runtime inference, not the trained data used to build the deployed model – results in a regression.

You also want to ensure that your code paths for generating features during training and during runtime inference are identical. This can, of course, be enforced at the code level – it should really be exactly the same code. Whether or not that can be enforced, parity can still be checked through testing. Testing for parity, while not exhaustive, can be relatively easily done by taking documents and running through the featurization in training and then in testing and simply testing to see if they are identical. This is an approach to testing that is also very useful when testing other forms of parity such as serialize/deserialize parity. It is worth calling out that to do this, we need to build on a basic engineering excellence practice – writing and testing equality between objects. With a correctly implemented and maintained equality implementation, writing parity tests in any scenario becomes almost trivial – simply generate two objects and test for equality. Implementing (and testing) serialization of features is also fundamentally important to the overall design of your pipeline as we will see later.

---

[4]Feature generation, or featurization, is the process in the ML pipeline by which raw input is converted into a set of fixed variables used as the input to training and prediction.

Now we come to the challenge of writing tests for a specific inference in your system. As we have discussed above, the models deployed in your code will change over time; and while an increasingly large core of your data population will enjoy stable predictions, there will be fluctuations in the margins as you continue to deliver net improvements. If we are building, say, a person name extractor, then your best model might no longer find all ten person names in document A even though the previous release of your model did. In addition, the component you are testing may be built on other components. For example, your person name extractor may be running after a subsystem that you have developed which identifies the main content in a web page (and discards the navigational, banner, footer, and other irrelevant content on the web page). This subsystem may also be based on statistical methods which exhibit the same sort of net gain progress.

There are several strategies involved in testing for this type of situation. First, construct a distinct regression testing framework which will allow you to run different models, diff the outputs, and inspect them. This should not be part of your unit testing suite as they aren't unit tests and they are long running tests which should be avoided in the inner loop. Second, different modules in the overall inference pipeline should be isolated. Rather than feed the output of module A into the input of module B in the test, the input to B should be the ideal output from module A. In other words, you test the performance of B assuming that module A delivered perfect output. Third, you should write tests for a specific version of a model. Rather than write a test for your person name extractor, you should write a test for version 3 of your person name extractor. This ensures that the test is deterministic and will always work in the same way. It will fail if something fundamentally changes in the implementation of your inference system, but it won't fail because you have improved your product with a net gain that has resulted in a regression of the specific case that you have tested.

We think of tests as pins that are holding up a large, strangely shaped picture on the wall. Put in one pin and fix one degree of freedom. Put in another and ensure that one flap of the shape remains fixed on the wall. The more pins put in, the more constrained the picture is until there are enough pins to guarantee that the picture will remain exactly where it is desired. The best tests are those that fail in completely unexpected ways, alerting you to code paths and input scenarios that you didn't anticipate.

Custom Labeling Tools

Labeling tools are the applications which humans use to create data sets for training and evaluation systems. They provide a view of the data as well as some controls that allow the user to enter or edit some sort of metadata. A simple example: If you are building a binary classifier for an image, then the tool might be showing the image and provide a button or instrument a keystroke to capture positive or negative judgments. We include the labeling tool in this chapter on technical and design excellence because understanding the purpose of these tools, the fidelity of the presentation, and the manner in which labels are created is important to the quality of your overall product. Let's break it down.

Viewing data might seem like a simple part of the overall puzzle. You are dealing with images, you need to view images; You are dealing with web pages, you need to view web pages. But it is worth digging in to the considerations in presenting data to humans.

First, are you showing the judge the same thing that your system is seeing? Certainly, when dealing with web data, it is easy to end up showing something to the judge that is quite different from what the machine will "see." If you save the HTML bytes of a web page to disc and then open them at some later date, your browser will start to pull in additional streams of information that it will use to render the page. This could be as trivial as the CSS styling information that determines the size of fonts, colors, and so on of elements on the page. It could also be imagery used on the page. Perhaps you are looking at a page relating to the news and there is a banner widget which pulls in latest headlines above the main content of the page. In addition, whenever additional data is used to change the look and content of the page, there is a chance that it is no longer present, resulting in unintended layouts, missing images, and so on.

Some of these presentational defects may not matter. But let's imagine you are building a classifier to determine if a page is "spammy" or not. If the junk content is no longer available to present to the user, then it won't be seen, and so the judgment can't be accurately made.

We have seen some more extreme cases of this type of problem when dealing with HTML data that is in some way restricted. For example, your data acquisition system might have access to servers hosted in a certain environment; but when you look at the same URL locally, as the permissions will be different for the individual user, the content of the page may be completely or partially blocked.

The web examples illustrate potential pitfalls in which the object being shown to the judge may be different from that seen by the system. We can also find examples where the human and the machine will see transformations of the data. It is very common for image processing systems to transform images to a common, lower-resolution version due to the capacity requirements for the training and inference system. In these situations, you have a choice as to showing the judge either the original source image or the reduced smaller image that the machine will see.

Storing and Versioning Training and Evaluation Data

There are two types of labeled data – the bountiful consequence of existing processes (e.g., queries in a search engine that resulted in either a click – success – or no click, failure) and the human-judged variety. The former is freely available, but the latter costs time and money. When working rapidly (not to be confused with agilely), it can be easy to complete your classifier, demonstrate its effectiveness, satisfy the customer, and move on to other tasks avoiding the overhead of persisting the data. When it comes time for you to update the model, the data's location and provenance are lost. As is the case with many processes in the agile world, the solution to this problem – this loss of value – is in setting up the right behaviors, processes, and tool chains that can be used *during* development. This means that at the end of the task – when your model is showing those net gains in the metric – there is no additional work; it's all taken care of.

The main considerations for managing data sets are the following:

- The population: Is the population an immutable set of objects? Any alternative means that you run the risk of high variance between samples. Let's imagine you have a store which is continuously changing – any time you take a sample, the nature of that sample will change. If this is by design, then well and good – perhaps you are building a new metric set every month. But if it is unintentional, then you will end up making assumptions regarding the equivalence of your samples.

- The sample: You can either sample by reference – in which case you create a list of references into your population – or sample by value, in which case you make a copy of the object data. There are pros and cons to each approach. If you sample by reference, then you ensure that there is only one unique version of the data point – it's the one in the population set. If you sample by value, then you have created a copy of the data which insulated you from any permanence issues your population data set might have. But now you have set up a situation in which the two objects may diverge for some reason.

- Versions of labels: When labeling data either for evaluation or for training and development purposes, you will want to revisit and correct labels. It is important that you keep versions of your labeled data. Why? Agility is about progress, and an efficiency in progress requires the ability to measure differences between experiments. It is very natural for anyone tracking the project to ask for comparisons that can be quantified, and you will certainly come up with new questions for models built on the latest round of labels that didn't occur to you several versions ago.

There are several practical approaches to managing data sets. For larger volumes, a basic strategy is to have some form of storage to manage your population and then handle smaller overlays of the data – references to subsets of the population that represent samples, labels associated with data objects – in the same infrastructure used to handle code. Git LFS (large file storage) is one example of a solution that follows this model. The basic workflow is to store your data set in the large file repository. Git LFS places references to the data in the main repository so that the Git interactions are seamless. A sample can now be defined by a reference to the objects in Git and label files (e.g., text files which reference a data object and serialize some label information) and handled in exactly the same way as code.

Managing Models

During development, and continuing as your product improves, the team will be producing many models and many versions of models that infer the same output. For example, when you launch your product, you may have arrived at version 7 of your steam engine name extractor. As you progress, you may create versions 8, 9, and 10 and arrive at version 11 when the next release goes out. Similarly, you may realize that the approach you were taking for that particular task was not ideal, and you want to go with a different type of model for the next round of experiments and release.

You have a choice when managing your models – you can either overwrite them or you can check in explicit instances of the model that can be referenced in the code. For example, if you encapsulate your name extractor in the class

`SteamEngineNameExtractor`, the overwrite approach would simply refactor this class with every iteration of the model. On the other hand, the explicit approach would create a class for each version: `SteamEngineNameExtractorV3`, `SteamEngineNameExtractorV4`, and so on. This latter approach allows you to explicitly run comparative experiments at any time between versions, allowing for both continual comparative metrics and on-demand comparisons with different data sets.

Compute as Much as Possible Inside the Tool Chain   Everything we do in the context of a data project can be characterized as

`inputData + (process + parameters) → outputData.`

Training is

`labeledData + (trainingAlgorithm + trainingParameters) → model.`

Inference is

`inputData + (modelRuntime + model) → prediction.`

With this in mind, we have a choice to make: how much of the process is run outside of our tool chain and how much inside? For example, we can train models outside our build framework and check in the parameters. This is a very reasonable thing to do if our training paradigm involves large data sets, lots of machines, and lots of time. However, if we are dealing with training systems that involve relatively small data sets and short training times, then there are advantages to training the model at build time. Fundamentally, it ensures that all of our systems are coherent. It guarantees that there are no bits of code known only to individual developers that are necessary to complete the training of the model. We can aggregate our metrics using a script saved only on our local machine; or we can have that script checked in, code reviewed, and tested. Even better, we can have our evaluation script executed with every check-in or automated to run every day. The more of our code we bring inside our process, the more efficient we will be; the more exercise code gets, the more likely we are to find problems with it and correct them.

Good Design for Data Projects

We've talked about technical excellence for data projects, so let's now move on to good design, in particular schema design – how to represent the input and output of your system. If you are building a system for analyzing PDF documents to deliver a logical document structure, what is the input – is it just the PDF? And what is the output – how do you design a suitable data model for your document? If you are building a system to extract business names from web pages, how do you represent the data at the end of the URL, and how do you deliver the business names found on the page?

We'll start at the end – what is the right schema for a business record? Table 9-1 shows a starting point, a simple string-based set of attributes. This looks reasonable initially, but let's consider some of the assumptions being made. First, we represent the address as a string. This poses a number of problems. Addresses have a reasonably well-understood structure. By representing them as a string, we are passing on to the consumer the task of figuring out what the structure is. There are three distinct things that we want to do with our addresses – display them to users, index them for search scenarios, and reason about them with respect to alternate forms and our data set as an aggregate. Displaying an address would seem like a scenario for which a string is perfectly adequate. But when your UX team wants to ensure that certain normalizations are guaranteed in the display component, you realize that even for that simple scenario, some internal structure would be preferred. Otherwise, you are asking the UX team to develop the domain knowledge required to normalize these strings – and that domain knowledge defines the value of your team!

Table 9-1. A simple business schema

Attribute	Type	Example
Name	String	Eric and Matt's Game Store
Address	String	1138 Lucas Street, Seattle, WA
Phone number	String	206 555 1970

So let's replace our address with a structured address object as shown in Table 9-2. In doing so, we need to acknowledge a requirement for data sources, producers, and consumers in our system – they have to be able to either generate or consume the new structured address format.

Table 9-2. *Refining the address structure*

Attribute	Type	Example		
Name	String	Eric and Matt's Game Store		
Address	StructuredAddress	BuildingNumber	String	1138
		StreetName	String	Lucas
		StreetType	Enum	Street
		City	String	Seattle
		State	Enum	WA
Phone number	String	206 555 1970		

Now we have a more useful representation of the business. We are thinking in more detail about the *decomposition* of the concepts being represented. Compare the idea of a string representing the address with having a structured object decomposing the concepts that go to make up the address. The string is simple and versatile, but it is opaque. The structured address brings value to the customer but requires more work. Of course, data is messy. A deeper review of the world of addresses reveals that not all locales conform to the street-oriented hierarchy common in European and Commonwealth countries. In Japan, for example, address structure is block based (lot number, block, city district); in some rural areas of many nations, addresses are more like location descriptions for the post office to decipher. This leads to a dilemma – do you attempt to create a universal data model for all addresses? Or do you incorporate a back off strategy that will allow at least some form of representation to be included when your structured object is insufficient?

An approach to back off representation is to permit a text representation in your data structure for any node in your hierarchy. In our address, for example, a further level of structure would be to bundle BuildingNumber, StreetName, and StreetType into a Street node. This node could have an additional property storing the string version of the decomposition. This would allow cases where the street wasn't decomposed or where it didn't fall into the model to at least have some record. Similarly, the top node in the model (the entire address) would have a text node which could hold the string form of the entire address. Table 9-3 shows the schema enriched with back off content.

Table 9-3. *Business schema with back off text nodes*

Attribute	Type	Example		
Name	String	Eric and Matt's Game Store		
Address	StructuredAddress	BuildingNumber	String	1138
		StreetName	String	Lucas
		StreetType	Enum	Street
		City	String	Seattle
		State	Enum	WA
Address text	String	1138 Lucas Street, Seattle, WA		
Phone number	String	206 555 1970		

Open Graph Representations   An approach to enable the modeling of structured but unknown schema is to have an open graph representation in which the names of the relationships (or parts) are undefined initially and represented simply as strings. This has the drawback of no compile time-type checking, but the advantage of being able to represent structured data with no a priori known schema.

Denotation and Identity in Data Models

Discussing data modeling leads us to the notion of identity. How do we determine if two records denote the same thing in the real world? To answer this, we need to firstly clearly define the concept that is being represented. What does our example record denote? It is easy to get overly philosophical about such things. So let's start with an exercise. Let's think about how we, as humans, talk about the business. We might say, "You should visit Eric and Matt's Game Store," to which you might reply, "Where is it?" At this point, you have hinted that the store is independent of its location. It could be anywhere (so the location doesn't define it). This gives us the idea that the **business** *operates* at, or out of, a **location.** But is this always true? What if we say, "Have you been to my favorite

Starbucks?" You might reply, "Which one is it?" The answer would obviously be a location. What we've stumbled upon here is the distinction between an aggregate concept like a chain or a franchise and a "singleton" business like Eric and Matt's Game Store.

This type of inspection, it turns out, can lead to fundamental insights regarding how to represent the core concepts in your domain. Our local search system had, for largely legacy reasons, a somewhat flat structure like that shown in Table 9-2. As we got more sophisticated with our project, we uncovered a variety of concepts:

- Singleton businesses: A business with only one location

- Business groups: Collections of differentiated businesses with a common owner – the primary example of this was a restaurant group in which a single company owned, for example, a sushi restaurant, a steak restaurant, a seafood restaurant, and so on

- Chains: A business with multiple of locations each of which offered the same service or experience (you expect to get the same burger at any McDonald's restaurant)

- Franchises

- Cooperatives

- Complex entities: Universities, hospitals, airports, governments, and so on

The biggest challenge that we faced wasn't understanding the structure of these concepts or even finding data sources that we could use in some way to populate a model with the required rich structure. It was the ability to migrate our schema from the original simple format to a structure that could represent the richness of the evolving model.

The strategy that we ended up adopting when it came to enriching our model was to think of the underlying, simple business representation as a base type and layer on complexity either as new attributes in the model or by creating additional models which captured new parts of the conceptual space and which could refer to other parts of the representation. For example, it is relatively straightforward to take the trivial schema shown earlier and add in a new field to represent, say, the category of the business. When dealing with chains, we created a new conceptual space of chain brands. The schema is shown in Table 9-4.

Table 9-4. *Chain brand schema*

Attribute		Example
Name	String	Josh's Poke Place
Locations	Array of references	[1, 88, 421]
Corporate phone number	Phone number	206 555 9943

Representing Ambiguity

Of course, the only thing we know for sure about our data is that it will contain errors. You will encounter enthusiastic declarations from the agents of particular sources of data stating that there are no errors in their data – don't believe them. We have found that even the business will not have accurate information about itself. Chains will include closed locations in their databases of active stores, phone numbers will have been left unchanged after an alteration, and so on. For you as a data designer, you may want to consider how you may represent ambiguity or confidence in your data. Deciding to do this is something of an important decision. The more you pass on ambiguity to your customers, the more you are asking them to apply some form of reasoning to your product to make it useful for their scenario. Is it useful to say, "Eric and Matt's Game Store is either at address A or address B – we just aren't sure"? Ultimately, someone, somewhere, must make a call. If you leave it to your customer, they will inevitably be forced to introduce an additional layer of reasoning to address the issues in your data resulting in uncomfortable dependencies and no real transactional trust. We have found that ensuring that your system has a well-designed mitigation path is just as important as the quality of your data. There will always be errors, but if you can immediately respond to those errors, you will get a lot of forgiveness. Designing for mitigation allows you to take a slightly stronger stance on the statements that you publish in your data – essentially you can make decisions, but you can fix them.

There are certain types of ambiguity that we feel are reasonable to pass along. Perhaps the most salient example is provided by the business name. While someone somewhere probably has a good idea of the unique name of a business, it is valuable to be able to capture two types of related names. The first is variations on the true name. Is it "Starbucks" or "Starbucks Coffee"? Is it "Home Depot" or "The Home Depot"? Capturing alternates like "Matt and Eric's Game Store" for "Eric and Matt's Game

Store" can also be useful. This type of alternate information can be further enhanced by including some indication of the distribution of use of the alternatives. This type of information is of particular importance in search scenarios. The second type of related information that is useful in our business listings scenario is associating the name of previous occupants with a particular store.

But let's catch ourselves. We are starting to go down the path of "attribute stuffing." We are adding attributes to our model to support specific customer scenarios. What is at the heart of this scenario? In recognizing the value of recording former occupants of a location, we have uncovered some additional key concepts. Firstly, that businesses are temporal. It is accurate to say that between date 1 and date 2, Eric and Matt's Game Store occupied a specific location. It is also accurate to say that at that time, the phone number which could be used to contact them at that location was phone number 1. The current occupant of the business unit at that location is Josh's Poke Place. There is an opportunity here to decompose the model and link statements in such a way that a consumer could compile a suitable representation for their scenario trivially from your factored model. This is the same principle that we discussed in defining and measuring the data product of your system. It is better to err on the side of the factors of the world you are describing than on the specifics of a particular customer or scenario. This is the analog in the data world of good class design in object-oriented software.

Representing Input

Thus far, we have discussed the data model of the output of our system. Inference systems generally deal with moving between different levels of representation. For our web extraction project, we start with some object data – a document – and end up with a semantic model of the content of that document. When we started the project, we used raw HTML data downloaded from the web site of the business as our input. This is a pretty common approach for applications that do classification or entity extraction from the Web. However, as we progressed with the project, we observed a couple of issues. First was the time of the download. This is important as without it we can't accurately track when an extracted field has been updated. Second, we noticed that we would sometimes either extract data that was not visible on the page and, at other times, not extract data that we would find on the page. In the first case, the problem was that we were treating all text nodes the same – even if the HTML was designed to hide some text from the user. In the second case, what the user saw when visiting the page in the browser was composed from not just the HTML byte stream that was downloaded when they navigated to that page, but also the CSS used to style the page, content dynamically loaded while the page was rendering, and modifications to the page that were carried out by scripts.

The temporal aspect of the data is an example of storing the *context*, or conditions of data acquisition. This data in and of itself constitutes valuable information for the overall system, and it is also useful for managing data sets – imagine a data set of images with the additional information of the time of day and location of the picture. The rendering issues which resulted in over- and underextraction of content are examples of fundamentally being able to describe the requirements of the system. While the general idea of the project was to extract business data from web pages, the implementation initially attempted to extract business data from only a partial and imperfect view of the data used to present the web page to the user. As we looked harder at this problem, we determined two slightly different scenarios. The first was to extract business data from web pages by emulating a person looking at those pages. The second was to reconstitute the database underlying the listing pages on the web site of a chain or other large, multi-location business. Thinking in those terms had a considerable impact on the abstractions used in our system as well as the design and implementation of extractors.

Conclusion

It might be fair to say that the most common confusion about agile engineering practices is the idea that agility involves cutting corners and avoiding tasks that don't appear to be on a direct line to the stated objective. But like any endeavor, efficiency of execution comes from a solid foundation in the basics. This is true of the foundational skills of the individual engineer, but also true of the foundations on which a team runs and the foundations on which a project operates. Early and continuous investment in quality will drive efficiency throughout the project.

In Chapter 9, "Technical Excellence", we have motivated the investment in excellence in software engineering in general – *optimize for developer productivity* – and discussed the parallels in data projects – *maximize the value of data.*

In Chapter 10, "Simplicity", we will describe how having a keen eye every day for work that should and shouldn't be done to move the project forward can make all the difference in a project.

CHAPTER 10

Simplicity

*Simplicity – the art of **maximizing the amount of work not done** - is essential.*

—agilemanifesto.org/principles

Agile development emerged in part as a reaction to top-down, committee-driven approaches to software engineering. At the time the agile manifesto was written, a number of innovations were starting to get traction in the development community, including eXtreme programming and Scrum, while various existing methods, such as Kanban, were being applied to the new faster-paced world of software development. Agile methods move the focus of problem solving from the upstream committee discussion to the in-flow engineering activity. Principle 10 underlines the reaction to this shift of focus, advocating the protection of the engineering capacity from the distant influence of the waterfall, putting decisions in the moment where the greatest relevant context can be found. As the old-school approaches are all but gone, the original motivation for this principle is less of a problem. However, it is still an essential skill of productive developers and teams to understand how to optimize the effect of their contribution by being selective of the work they do in the interests of the project. In addition, the dialogue that emerges from the permission to be selective is an excellent way to hone the focus of work and scope the project.

The terms used in this principle are a little too open to interpretation. Rather than worrying about what "simplicity" means with respect to engineering processes, let's consider the spirit of the principle.[1] The principle is not about avoiding work, but

---

[1]Steve Jobs said, "Simple can be harder than complex: You have to work hard to get your thinking clean to make it simple. But it's worth it in the end because once you get there, you can move mountains." On the other hand, Benoît Mandelbrot said, "To simplify, complexify." Jobs was talking about reducing the dimensions in a solution to achieve elegance; Mandelbrot was talking about introducing degrees of freedom to better understand and manipulate the problem space.

© Eric Carter, Matthew Hurst 2019
E. Carter and M. Hurst, *Agile Machine Learning*, https://doi.org/10.1007/978-1-4842-5107-2_10

rather ensuring that the work being done is intentionally and clearly driving the project forward. This type of focus is an important part of the feedback loop between doing things and deciding what things to do.

Being Diligent with Task Descriptions

In scrum, much of the planning and decision-making process culminates in writing user stories and task descriptions. These provide a useful inflection point in the ongoing project process where the potential for embarking on unnecessary work can be checked. A few straightforward tests can be applied to the language used to describe planned work that will help avoid some obvious less productive investments.

Underspecified Work

Underspecified tasks can lead to work that doesn't move things forward. A frequent example we've found that crops up in data projects is the *improvement* task: "improve the person name extractor," "improve the business page classifier," or "improve the regression testing framework." This example gets at the fundamentals of managing data projects – balancing the unknowns in the data and models with the knowns of discrete chunks of estimable work. A metric goal is achieved by a series of tasks that explore the search space of possibilities in developing a model. For example, a gain of ten points of precision with no regression in recall for a classifier might be achieved after running 20 experiments that involve cleaning training data, parameter sweeps, additional features implementation, adding the ability to view false negatives in the data viewer, and so on. A minority of these experiments will be on the ultimate path that leads to the desired metric goal. Adding features to the tool chain will result in further insights that will allow for the refining of the plan. Each of the tasks involved in setting up and running an experiment, or building up features in the tool chain, can be managed in the traditional manner with costing being estimated with reasonable accuracy. If the team could perfectly plan a single path through this search space of data and models, then achieving the goal would be done in a predictable set of steps. However, this is, of course, impossible. And so, the team has to use some general strategy and intuition to prune the search space as it progresses toward the goal. This is done in the iterations of planning for the project – the sprint. Distinguishing goals from tasks gets you to a point where you can actually plan work and manage the team's resources. Layering the metric on top of this activity with short iterations and frequent experiments allows you to see the trends in progress. This will give you a handle on estimating to some degree the effort required to get to a metric goal.

The word "improve" indicates two conditions – first, that the component is not as good as desired and, second, that there is no clarity on how good it needs to be. Any developer given this task should first push back and ask, "What is the current state? What is the desired state once this task is complete?" In addition, even with clearly defined metric goals, better tasks can be written that describe exactly the work that is to be done – the application of the strategy, not the goal of the work. Any task that can't be estimated should be decomposed into either tasks that can or tasks that are aimed at reducing unknowns that prevent later tasks from being properly estimated.

If the improvement is intended to address a specific error case, for example, an area of the data population where the classifier systematically does poorer than in general, the task description can be rewritten. Instead of "improve," start talking about increasing the metric (precision, recall, etc.) for that well-defined case or subpopulation, and then decompose to discrete tasks. This will naturally lead to subtasks involving sampling the population for this case, measurement, and analysis, as well as the experiments through the model space – new feature extractors, additional data labeling, label correction, and so on.

Now that you know what you are working on and the starting point, what about the target? There are a number of ways to gain clarity here. Firstly, consider the role of the improvement. While the most obvious case is to deliver value, you might also find yourself working on a phase of the project where the goal is to demonstrate capability. In this scenario, you are interested in gaining knowledge regarding the amount of effort required to see a meaningful change in the metric. Consequently, the target should be something that demonstrates potential. Think in terms of the error rate. Can you reduce it by 10%? If you are aiming to deliver value to the end customer, then you want to consider metric changes that are visible and meaningful to them.

If you are dealing with particularly hard problems, then it might be important to structure the work in terms of the trend of improvements. A 1% metric gain might not be something that your customer cares about, but showing a constant gain in the metric over time toward a meaningful goal is a great demonstration of capability.

In all these cases, where a metric is involved in setting targets or tracking progress, consider the power of the metric. Is the change statistically likely to have occurred given the sample size and quality of judgment? We have seen cases where progress toward a metric goal falters. Upon analysis, we've found that the noise in the measurement data (due to the complexity of the judgment task, the sophistication of the tools, the clarity of the guidelines, etc.) can swamp any improvement.

Without drilling down into the details of the "improvement" task, you run a real risk of taking on work that has no observable completion. You'll meander on, not knowing that you've already created value, without integrating the insights you've gained back into the priorities in your planning. Ironically, it may be wasteful to exceed in improving an inference component due to the opportunity cost of other higher-priority investments that the customer would prefer be made.

Deadly Conjunctions

In the previous section, we looked for the word "improve" in task descriptions. Another problem word is "and." If you see "and" in a task description, raise your red flag high. By conjoining tasks, say "analyze classifier errors and improve recall," you assume that the insights from the first won't impact your decision to do the second. In addition, you prevent the completion of tasks – an important behavior in keeping a nimble and motivated team running. You also run the risk of mixing estimates of cost for the tasks. An analysis, for example, should generally be timeboxed. Making improvements – as discussed above – brings its own complexities and unknowns. There is great value to the team in completing a task. Splitting conjoined tasks into two doubles the completion moments that you can celebrate.

Cross-Task Dependencies and Assumptions

Poorly worded task titles and descriptions can lead to miscommunication in the team, resulting in confusion about who is doing what and how individual pieces of work fit together. If you expect another developer to deliver something as per your interpretation of the task but in fact they are working on a different interpretation, then the work that either or both of you are doing may be heading in the wrong direction.

In a recent project, an agreement to produce and automate a metric computation on our predictions was split between two teams. Once both work items were complete, we discovered that the metric automation was delivered for a different, extant metric, while the results of the new metric computation were left untouched.

When participating in planning processes, if you are aware of assumptions you have about a particular task, it is best to voice them to help reduce potential for confusion and wasted effort down the line.

There are several things we have found that greatly help with chasing out dangerous assumptions:

- Planning poker: This is a planning game that uses the estimates for the cost of a task from two or more people to surface disconnects in what the task actually is. In summary, the players each give an estimate (without knowing the estimates of the other players), and the numbers are compared. If they are similar, then no further discussion is had. If they are quite different (e.g., one estimate is twice that of another), then a further round of discussion is carried out, digging deeper into the specifics of the task. It's a great tool to find these disconnects as it only requires a few moments thought to form a rough estimate, and this is exactly where assumptions will be used. For example, let's say the task is to build a simple labeling tool to annotate phrases in text. Your estimate is 2 days, but mine is 5. Our coding skills are not dissimilar, so why the difference? In discussing the task, it turns out that I'm assuming that we need to implement management for different labeler identities, while you are assuming that you will just write the labels to disc in an unadorned format.

- Pair programming: If the producer and the consumer can sit down together to implement the API, component, or data transformation, it can obviate the need for a more formal documentation process and allow each side of the exchange to quickly iterate to get to a final result that will have minimized assumptions on both sides.

- Synthetic data: Even before the API is created or any data has been processed, it is useful to manually (or through some simple automated process) produce some synthetic data. This unblocks the consumer from having to wait not only on the design but also the implementation before they can code against the data and discover issues with the data model or the interchange format.

Words Matter   We touch on communication as a core skill at a number of places in this book. Like good code, clearly articulated descriptions of work can prove crucial to the team's ability to do real work in a sprint. Developing a critical attitude to the quality of the language that is used around your project can help you navigate the landscape of data projects.

Early Integration

A common decomposition in large projects with multiple ML components is to work toward metric targets at the integration points. Your team will produce data at 90% precision and 80% recall, and my team will produce a search ranking function with a ten-point improvement of DCG[2] over the current ranker. In reality, it is difficult to predict what the upstream team needs to deliver in order for the downstream team to build on that improvement to deliver their goals. Consequently, the downstream team is motivated to request a metric goal that is safe for them – in other words, it is likely to be higher than required.

In an ideal world, you would ensure that the two systems are integrated as soon as possible. In addition, the downstream system would be designed to support low-cost experimentation. By doing this, the metric target at the point of integration becomes less of a concern – you can directly observe the effects of the upstream improvements on the final product. In comparison with the more decoupled approach, this provides the opportunity to complete the project earlier.

Baselines and Heuristics

Given a new problem, it is common to see an inexperienced team dive in with the latest machine learning methods, eager to show their chops – they are asking the question "What happens when we apply this method to this data?" Complimentary to the importance of establishing metrics, coming up with some sort of baseline system can deliver useful insights for very little investment. This can be as simple a method which always outputs the positive class for a binary classifier. More informative is an

---

[2]Discounted Cumulative Gain – a metric for measuring the accuracy of a predicted ranking with a known, correct ranking.

implementation of a small number of "obvious" rules capturing your intuitions and assumptions about the problem. For a little more effort, you could train a simple logistic regression model as a baseline for more sophisticated ML methods. By coding these up and evaluating them against your metric, you will learn something about the complexity of the problem space. Furthermore, by looking at the errors your simple heuristics make against the metric set, you will get a feel for the variance in the population.

In our system for extracting business listings from web pages, we were well set up to support a number of approaches to entity extraction using machine-learned methods. We supported both sequential models and classification models. However, when it came to deliver a phone number extractor, we started with a simple regular expression. As the project progressed, we learned more about the patterns for legal phone numbers in different countries (e.g., in the United States, not all sequences of ten digits are valid phone numbers, which is why numbers in films always include a 555). The combination of this knowledge and the pattern-based approach was sufficient for extracting phone numbers at the required level of precision and recall so we didn't even invest in training a model.

In comparison, our investment in address extraction started with a legacy pattern-based approach using a finite state transducer. While this approach was convenient in that it already existed, we soon found its limits and switched to a full ML approach.

Recognizing Limits

As data scientists, we are keen to remind people of two universal truths: the data is always noisy, and no inference system is perfect. With the latter, we are setting the expectation that there will always be errors in the output. We have talked in Chapter 9: Technical Excellence about ensuring that your system is designed with mitigation as a primary capability. This allows you to react to errors as they are discovered and fix them. This is great for addressing highly visible and potentially embarrassing or dangerous errors. However, it is an after-the-fact solution. The error was already exposed.

Most large-scale, ML-based engineering teams will have already built systems around crowdsourcing data. These involve data and job management systems as well as tools and processes for training and managing pools of remote or on-site judges. These judges will either select or be assigned work which they perform for a small unit price. This type of pool of human intelligence can be integrated into your data processing workflow – we call this the human-in-the-loop model.

In the world of local search, all records appear the same: the name, address, phone number, and so on of the business or local entity. However, depending on the scenario that your data is used for, the importance of the entities can vary greatly. In fact, in search, a large percentage of entities get shown to any user only a handful of times and some percentage never at all. Consequently, the quality of your system is more sensitive to errors in the important, hero entities[3] than it is to those entities that never show their face. With this in mind, we want to make sure that changes the system infers to those hero entities are not only more likely to be accurate but are actually verified by a person. It would be very unlikely that the address of the White House changed, but if it did, you don't want to be embarrassed by not reflecting it in your product.

To handle this through the human-in-the-loop method, we identified a set of entities that we would monitor. Whenever a change was computed for those entities, rather than allow the change to carry through the system as with other entities, we take them on a little detour. They are marshalled for review by human judges. If the changes are verified, then the information simply flows back through the system. If not, the error is not only discarded but recorded so that the system doesn't fail for the same problem again. For the cost of a little latency, we can ensure that important entities are protected.

The nice thing about this approach is that it scales easily to the capacity of your human judges. You can simply rank entities by their importance as reflected in the number of times they are presented to users. Whenever a judge is free, they can review the next entity on the list that has had changes in the current cycle.

By designing for a human-in-the-loop, you avoid the increasingly difficult and expensive task of finding ML-driven approaches to solving the last few points of quality in your system. In addition, this avoids the increased expense of funding a metric system that is sensitive enough to even measure with any accuracy changes at that level.

Managing HiPPOs

The further away someone is from your project, the less context they have about the trials and tribulations of the developers in the trenches and the less understanding they have about what is easy and what is hard in the problem space. We have all had the experience of looking from the outside in on a problem space that another team

---

[3]The term "hero" is often used to indicate entities, results, or scenarios in data-driven projects that are too important to fail.

is attacking and thinking – "how hard can it be?" When the person with this view is somewhere above you in the management chain, it can be an easy choice to make to react to their input immediately, breaking the sprint, throwing planning out the window, and generally rushing around. It is important to be aware of the impact that these people can have – we call their input HiPPO (Highest Paid Person's Opinion).

There are two general approaches that are worth considering. First is a strategy we might call transfer of decision. When asked by someone with influence to take some action, you should respond by asking them which of the current priorities should be changed. This is an effective way of handling the situation as it provides an opportunity for education and also makes the consequences of any sort of veto power clear. Of course, this can only be done well if you have a well-managed backlog and sprint plan. If you are running by the seat of your pants with ad hoc, just in time planning "methods," then you are wide open to vicarious course correction.

The second approach is to rely on data. If asked to do X instead of Y, you can respond by either referring to existing data or doing some quick, but principled, analysis. If the insight of the HiPPO was valid, then good. You have been nudged to better understand your data.

In both of these approaches, the key investment is what happens *before* the intervention. Your planning should be worked out (in the scrum sense, not the waterfall sense), and your metrics and data should be mature enough to facilitate informed discussion. In general, we find it useful to be one step ahead of any potential interlocutor of our data and metrics. If your metrics are going to be used in any interaction or some analysis of your data set, having the next level of detail, the distribution of subtypes, the breakdown of error cases, and so on in your pocket will go a long way to ensuring that the right decisions are made efficiently.

Failing Fast

Agile development for data projects fundamentally changes that landscape of risk in comparison to traditional engineering projects. The space of options is compounded by the unknowns in the data and the challenge of the inference problem at hand. It is reasonable to assume that the paths to success are reasonably limited given the dimensionality of the data. Consequently, the ability to prune the search space expediently is fundamental.

Fail fast[4] is a strategy by which you orient your work to determine as rapidly as possible if the general approach is viable or not. You want to demonstrate that the approach will fail as quickly as possible to avoid longer investment in a solution that will not work. The corollary is, of course, that if you fail to demonstrate this, then the approach under consideration may well have legs.

General considerations around fail fast include the following:

- Understanding and testing your assumptions about the component or problem: Work hard to surface assumptions in your plan – which things must necessarily work well for you to deliver. What evidence do you have that they will? You are almost certainly applying the method in a new scenario, or to a different type of data.

- Scale: What happens to the component when it has to scale? It is not necessary to implement scaling, but you should certainly look for opportunities to figure out the key aspects of a scaling strategy and do some quick testing, possibly with synthetic data or simulated workloads.

- Labeled data generation: Is there a path to synthesize or otherwise automate the collection or creation of data? If you require human judges, is it a task that can be accomplished with your current tools? We've found that if you do some quick labeling yourself, you might learn that the central concept of the problem you are trying to solve is actually quite vague and may run the risk of low-quality labeled data production.

Build or Buy or Open Source

Strategies addressing the amount of ownership a team has in a solution define the identity of the team. On the one hand, a complete ownership approach establishes the team as a center of excellence for that problem space. At the other extreme, a strategy that is committed to integrating external components and data sources identifies the

---

[4]Fail fast when used to describe a strategy for system design and implementation is actually a metaphor referring to the fail fast approach to implementation, in which code is designed to raise exceptions, or otherwise fail, as early as possible to avoid wasted time and computation.

team as a platform and integration team, potentially with a considerable requirement for business relationships. It is important to keep this in mind as it will influence the type of team member that you will want to attract.

There are several ways in which the buy vs. build consideration comes in to play when your team is tasked with delivering data products:

- Buying or licensing completed data sets and the services to update and maintain them

- Buying data sets that will form a part of your solution, which you will improve and augment

- Buying inference components that you can apply as is to the raw input data that you acquire yourself

- Buying tool chains that will support your team in developing the models that are required for the delivery of your product

In all these scenarios, the key thing to consider is how agility will be managed through the dependency. For example, how will DSATs[5] be reacted to? How will extraction errors be addressed in the extraction software? How will buckets of false positives be prioritized and improved upon in a classifier?

If you take a full dependency on a data source, there will be an expectation that prioritized problems and individual errors will be reported to the source of the data and addressed and then a new version of the data will be published with the expected changes. A third-party data source is of interest because the provider has more resources to put into the data. The dependency makes economic sense because the costs are shared across multiple customers. However, this reduces your ability to influence the priorities of issues being addressed. Ideally, the third-party provider will have a mechanism that will allow you to post reports of problems or actual data corrections. Potentially, the validation and application of corrections can be part of the service – you might pay per correction.

The alternative to this is to apply patches to data on your side. This is essentially the conflation scenario – the third-party data is a signal which is mixed with at least your corrections data and potentially other third-party data sets.

---

[5]A DSAT (antonym: SAT) is a customer experience that is dissatisfying. It is generally styled as a capitalized word.

In summary, a data dependency is either a strategy that is wholly dependent on the latency and prioritization of error correction on the side of the service provider or a strategy that requires investment in conflation and correction data management on your side.

When it comes to dependencies on third-party inference components, there are similar considerations – how to manage improvements to the system to address buckets of errors and specific DSATs. Improvements on your side can be managed by wrapping the third-party component and applying appropriate lightweight rules to correct specific inference errors.

Establishing an Integration Framework   Something that we realized after a number of projects involving a mixture of first-party and third-party components was that rather than thinking of the third-party selection and integration decision as a one-off consideration resulting in specific designs around integrating that specific implementation, it was better to think of the problem as providing a general framework for integrating any number of potential solutions to the problem. This approach has the benefit of allowing you to swap solutions in and out as needed, but also it allows you to mix and match solutions as they apply to different areas of the problem space.

For example, if you are evaluating solutions for document understanding – that is, inference components that determine the logical structure of a document given its raw input – you might find that one solution is good at general high-level layout while another is good specifically at recognizing and structuring tables. You want to have the advantage of the latter in the context of the general value of the former. To do this, you can build a framework within which both can run. The framework will then compose the results into the final output.

Conclusion

In Chapter 10, "Simplicity", we have introduced methods for ensuring that individuals and teams can focus and act efficiently through the planning and execution of well-considered work. You are a member of, or a manager of, a team of smart data developers eager to make an impact on the world through your services and data products. You should come in to work every day with a keen eye for relevant work that is aimed at moving a project forward. You should challenge your customers to be engaged and present in the progress of the work. Within the team, you should help guide the progress through the unknown and un-navigated landscape of your data problem by helping to scope the problem, look for knowledge gaps, and articulate well-defined experiments.

In Chapter 11, "Self-Organizing Teams", we will describe the personalities that can be found in many teams working on data projects and how to leverage diversity in the team to produce a better product.

CHAPTER 11

Self-Organizing Teams

*The **best architectures**, **requirements**, and **designs** emerge from **self-organizing teams**.*

—agilemanifesto.org/principles

Top-down planning processes and hierarchical team control impart a certain kind of comfort to management. This warm feeling, however, is only really experienced at the top and rarely by those executing the actual work at the bottom. There is nothing too mystical about this principle – on agile teams, the low-latency communication within the team combined with the license to experiment, make mistakes, and course correct allows the design of components, APIs, data structures, and so on to be just what is needed and just in time for the task at hand. Sometimes what is happening within the team isn't as apparent to management, although the techniques discussed in Chapter 4: Aligning with the Business go a long way toward helping management regain the warm feelings they lost when waterfall went away.

In the context of data projects, teams are often formed around areas of data, and so team members are the most experienced and knowledgeable about the intimate details of that data – far more than any external person or person higher up the management chain. As the data is a huge factor in what the team can accomplish and how it accomplishes it, having the data experts be the primary voice in all aspects of team process, planning, and execution makes a lot of sense.

In *Empirical Findings in Agile Methods*, Lindvall et al.[1] describe agile methods as: **iterative**, **incremental**, **self-organizing**, and **emergent**. *Self-organizing* they define as: the team has the autonomy to organize itself to best complete the work items; *emergent* as: technology and requirements are allowed to emerge through the product development cycle. Both definitions point to the absence of top-down approaches to

---

[1] www.cs.umd.edu/~mvz/pub/agile.pdf

E. Carter and M. Hurst, *Agile Machine Learning*, https://doi.org/10.1007/978-1-4842-5107-2_11

planning, design, and execution; but they don't imply that the team determines the high-level tasks or goals. A self-organizing team is not a team with complete autonomy. At some level, projects, targets, and business necessarily will be determined by some external agency. What is inherent in self-organization is autonomy over *how* something is carried out. Furthermore, a team cannot function without process. Process *enables* productivity. This can be hard for some to grasp as they may perceive process as a hindrance – an unnecessary layer of administration sitting above the more important work of writing code and training models. Managers of data teams must figure out how to enable their team via processes that work.

Team Compositions

In projects at Bing and elsewhere, we have observed several types of team composition, all with different pros and cons. These team composition types include **balanced capabilities, mixed capabilities, stratified capabilities**, and the **surgeon team**.

Balanced capabilities occur when members of a team are, for all intents and purposes, interchangeable. There is no real difference in outcome if individual A or individual B picks up a task. It is very unusual to have a team composed of just these individuals; it is also inadvisable. Given that you want to be able to cover the full stack, you would need to populate your team with the mythical full-stack data scientist – an almost impossible task. It is important to recognize this when building teams and in the planning process for sprints as well as the day to day running of the team.

Mixed capabilities are found in a team of equally adept but complimentary developers. This is perhaps the most effective team composition to have. There will be some individuals with deeper data science expertise, some with more of a backend engineering focus, and so on. Self- and group-learning will happen if the team members are passionate in their field and supportive of each other. The key thing with mixed capability teams is to ensure that knowledge and know-how doesn't get trapped within a silo of individuals with similar capabilities and is constantly socialized within the team, especially with individuals with different capabilities. For example, the data scientist focused team members should regularly pair with the backend focused team members – either by coding side by side or at least sharing design and architecture insight and ideas with one another. Specialized capabilities can produce great results and great dynamics in the team, but hidden know-how and tools can lead to disaster and lack of productivity, especially when someone leaves. And often the best ideas come when individuals with different capabilities work together on the same problem – a data scientist can often have an insight that the backend engineering team hasn't considered and vice versa.

Stratified capabilities are found generally when the team contains junior and more senior individuals. Again, this a very normal and constructive type of difference to have in a team. Senior individuals can be supported by more junior members who gain experience from the interaction. In addition, in the fast-paced world of machine learning and data science, new college hires in particular are great conveyors of new norms in how particular problems are being solved. They can help to shake things up in a positive way.

The surgeon team is centered on a small set of high-performing individuals. The surgeons are like the surgeon in an operating theater – the main conduit through which progress is made and the person whom others pay attention to and generally follow. This can be an extremely difficult team to effectively land results in a data engineering project. Too much is centered on one individual, and in typical data engineering projects, the data is too dense for one individual to fully consider. While the surgeons may be highly capable, their centrality in the team process can block the team as a whole from reaching its full potential.

Teams Are Made of Individuals

Part of the fun of working in a team is the spectrum of capabilities and personalities that you will work with. We now consider some examples – caricatures to some extent – of types of individuals that often show up in data-oriented projects: the **new hire,** the **journeyman engineer,** the **philosopher engineer,** the **principled applied scientist,** the **surgeon,** the **minimalist,** and the **off-roader.**

The *new hire,* fresh from graduate school with a PhD in a machine learning- or an AI-related area, is a great source of new methods, new perspectives, and energy. This team member has been living in a world of cutting-edge techniques, unencumbered by the pragmatic friction of the industry, and has been optimizing for solutions to complex problems rather than the engineering matrix required to ship products. They bring fresh insights and can challenge the team. They often have little experience in production quality code, so they need to fully embrace software engineering practices as dictated by the team's culture. In an open and welcoming team, you should find many candidates interested in mentoring new hires like this.

The more senior *journeyman engineer* is a solid, dependable contributor. They are perhaps less experienced in machine learning methods but make a large and important contribution in ensuring that designs take into consideration production concerns such as scaling, testing, and deployment automation. They are very experienced in collaboration, especially with complimentary contributors who provide capabilities in the overall system that are outside the journeyman's purview. They may be interested in some serious career development through learning about ML, but even if not, they are invaluable to the team.

The *philosopher engineer* is a little less motivated by the pace of delivery and the business behind the project, but is very motivated by ensuring that data models, architectures, planning, and any other declarative aspect of the project are done "right." They have a powerful sense of the natural and intuitive decomposition of problems (often in a manner independent of the computationally optimal approach). They will balk at any sign of an architecture shaped by the organizational structure rather than the nature of the problem. Their palms will sweat when they see data structures with overloaded semantics. They may get hung up on design and need to learn to balance their interests against the required velocity of the team. They may well make principled but unlooked for investments in refactoring code that doesn't sit well with them, and so managers need to think about when and how to encourage this or channel the energy elsewhere. The sustainable pace ideas in Chapter 8: Sustainable Development can help with this.

The principled *applied scientist* is a little like the philosopher engineer – they care a bit more about doing things right than other product considerations – but their obsession is the details of machine learning. They are invaluable in ensuring that all the small details of machine learning are taken care of so that the team gets the most out of their accumulated wisdom. Junior and less experienced ML practitioners would do well to listen to them to pick up a broad spectrum of tricks of the trade: how to clean and balance data sets, how to normalize feature values, how to analyze outliers for clues to classifier improvements, and so on.

The *surgeon* is a highly productive, charismatic, and principled developer who has an inordinate influence on the team primarily due to the velocity at which they work.[2] By being productive, they become involved in many of the components in the product. They are a blessing in that they keep things moving (sometimes at an alarming rate), but they may also introduce challenges in that they can accumulate bottlenecks and to a large extent dictate what other individuals should be doing. In the best cases, there is a clear net win for the team being generated by a surgeon. In the worst cases, their negative influence is greater than the benefit of their expertise and the team ends up being herded down a path that is more to do with force of will and less to do with principles.

The *minimalist* finds themselves in a team with burgeoning opportunities and more work than can be carried out effectively by the group. They become focused on ensuring that the work assigned to them is completed, but are reluctant to stray into unknown areas or take on problems which they don't "own." They lack the natural exploratory nature of a data scientist and don't feel empowered to take on unknowns where they might fail. They like well-structured work where the accountability is externalized.

The *off-roader* can't stop prototyping. They often show up with fully functional prototypes in areas that were not originally central to the thrust of the project but which, by their mere existence, influence the direction of the team. They tend to ask for forgiveness not permission. They don't feel limited by "official" statements of vision or mission and through their enthusiasm will contribute to both. Their prototypes can give a sense of possible directions to the team, but must be carefully managed as the requirements put on a prototype often don't represent the real-world conditions of the product. Off-roaders can often generate prototypes that don't scale well when actually exposed to real data.

Individual Traits to Encourage in a Team

From these caricatures of individuals, we can pull out some key traits to be encouraged: **communication, relating the small to the big, keeping an open mind,** and **equitable participation.**

Communication is key: A good communicator knows how to express current state, how to participate in a constructive group discussion, and how to get information out of others no matter what others' communication capabilities might be. They know

---

[2]C.f. the 10× programmer from Exploratory Experimental Studies Comparing Online and Offline Programming Performance (Sackman, Erikson, and Grant, 1968).

how communication works within the team, across teams, and up the management chain. A good communicator understands that clarity is a key leadership quality and demonstrates this at a personal level while expecting it from others, especially those in leadership positions. They nurture the skill of communication within themselves and within the team.

Relating the small to the big: This trait leads to sampling data to understand the population, writing unit tests for even apparently trivial methods to ensure that the code's foundations are strong, caring about the details while keeping the big picture in focus, and helping team members be accountable. An elevated position or a leadership role shouldn't mean that only high-level, high-scale concepts are considered.

Keeping an open mind: When CEO Satya Nadella began his work on transforming Microsoft's culture, he tapped into the idea of a "growth mindset."[3] Having a growth mindset means you are able to listen, you are able to consider things from another's point of view, and you are aware of how assumptions influence your thinking and decision making. Without this key attitude, teams can become dysfunctional. Opinions become entrenched, trust is lost, and morale suffers.

Equitable participation: A self-organizing team, to a large extent, should be self-sufficient. This means that a complete spectrum of work needs to be handled. With a growth mindset, this leads to opportunities to mentor (you can help me understand a part of the system that I'm less familiar with), motivation to build productivity tools to support repetitive tasks (you and I both dislike doing something, so we can agree that building a tool to automate it is a good investment of our time and will make us more productive in the future), and easier interactions (I don't need to feel bad pointing out that you haven't been pulling your weight in a certain area). Managers don't want to have to remind begrudging team members to participate, and colleagues don't want to experience awkward meetings where their peers are shamed.

There will always be outliers in the interests and passions of your team. In some cases, these will be complementary. Matt really enjoys labeling data and building

---

[3]Growth mind-set is a term created by Carol Dweck, a professor of psychology at Stanford University, and described in her book *Mindset: The new Psychology of Success.*

tools to make viewing, interacting, and annotating data easier, for example, but has found that there are many who are less excited about this – and that's fine. Many guides to agile processes assume that all team members are fungible. In reality, this is not the case (of course). What is important is for the team to not have vital information, code, and know-how locked up in a single engineer's head. A good autonomous team will ensure this doesn't happen. Behaviors that help guard against this sort of knowledge isolation generally socialize the code and workflows involved. These include the following:

- Reviewing and checking in code in the open so that others have a chance to look at the code and ideally to understand its design. Azure DevOps provides great tools for doing code reviews on pull requests so that everyone can comment on and view code going into the system.

- Ensuring that all code is built on the server so that there are no peculiar configurations or other dependencies local to the individual's setup that prevent others from easily using the code. This was discussed in more detail in Chapter 3: Continuous Delivery.

- Evangelizing, demonstrating, and supporting code, tools, and processes will provide a trial by fire. If no one can be persuaded to use something, then perhaps it doesn't yet fill a clear need.

Managing Across Multiple Self-Organizing Teams

Bing's local search was comprised of several teams all following agile processes and executing in a self-organizing manner. We found that there were areas where it was advantageous to institute shared practices and tools as well as assign certain tasks across the broader pool of engineers from the combined teams. Some of these requirements can leverage a central tool, for example, if a certain level of code coverage for unit tests is required, there is no need to invent a new way for each team to measure this.

As the teams were collectively responsible for the product, the basic life cycle elements that enable progress and underpin the maintenance of the production system were at least known to all team members. This basic knowledge helps keep a general sensibility about the product present in all components, but also means that any individual can contribute to any number of failures. For example, a common tool for immediately addressing errors in the data is available to any team member and allows them to correct any data that is formally or informally reported. This type of knowledge – including deployment, rollback, maintaining the build, ensuring the integrity of the code base, and so on – is necessary for the DRI[4] role.

Coding best practices and style requirements are enforced primarily through the use of specific tools. Coding productivity tools – in this case, ReSharper – are used in the IDE, while other code policy checking systems were deployed to prevent code being checked in if it didn't conform to certain rules. These stylistic policies are important when it comes to sharing and reusing code across teams, as well as supporting future engineers who have to maintain the system long after the original author has left. Code should not only be functional; it should be *understandable*. The code itself can be written in an opaque manner (abstract variable names, uninformative method names, poor formatting, long lines, lack of namespace structure, etc.) and worse, lack in comments and documentation. The best code would require no documentation but would be "self-documenting." That isn't always possible, and so paying attention to the details of style will help later engineers understand. Code style adds future value.

Empowered Teams Drive Team Development and Product Evolution

Our experience in the local search team at Bing led us to some useful insights regarding the value of autonomy within teams especially as it relates to data-intensive projects.

Data projects continuously *surface new knowledge* regarding the problem space, the data, and the efficacy of particular approaches to delivering on a target. It is

---

[4]Designated Responsible Individual: It is common practice in teams responsible for live production code such as that running an online search engine or other services to rotate members through this role which requires that they be on call in case of critical system outages and other problems impacting customers.

vital that this information is effectively communicated outside the team to ensure that progress is transparent. Metrics should be tested for how easily they can be communicated – do they make intuitive sense to those to whom you are going to report? Iterations should be short enough so that a trend can be generated – this is your key expression of progress and a useful way to help others get a sense of when something will be completed.

Discoveries relating to the data being processed and the particular goal of the project can provide a great source of *ideas for new features.* Teams can champion these data-driven insights and help drive innovation at the higher product level. When working on delivering hours of operation for local businesses, we learned a lot about the frequency with which businesses changed their hours (as indicated by the changing hours on their web sites). While an individual business doesn't change its hours too often, at any given time a significant number of businesses will have changed their hours within a country. In addition, a number of chain businesses in specific consumer areas such as DIY and tax services will change the hours of all their stores according to a seasonal schedule. When comparing what we were extracting from the Web against the alternate signals we were getting from traditional sources, we found discrepancies. Sometimes a business would change the hours on their site before updating their data through other channels. In other cases, they would change their hours first through their SEO channels and then on the Web. It occurred to us that there was an opportunity to inform the businesses of these discrepancies and ask them which version of their hours was correct. This was proposed as a feature and, in collaboration with a partner team, delivered to the Bing experience for small business owners.

Active, passionate teams can be *self-educating* – organizing paper reading groups, participating in online courses *en masse,* and presenting and attending data-oriented deep dives or brownbags on relevant topics of interest. This interest in education is incredibly important. It's not a simple matter of self-improvement, but a great way for the team to keep abreast of new research, open source models, and so on. The passion motivating these types of activities often comes from individuals, but the license to do them comes from the culture of a self-organizing team.

Teams evolve their own operating culture influenced primarily by the members of the team and their values and their experience and comfort with the various tools and processes available to support agile engineering. There seems to be a very noticeable difference between teams of three to five individuals and teams of five to ten individuals. In the smaller team, the individual in the role of managing the team is almost certainly also operating as a line engineer: writing code, reviewing code, and so on just like everyone else. Consequently, it may be attractive to ditch as much process "overhead" as possible as it gets in the way of coding. However, in slightly larger teams, the manager may be playing a less critical role as a developer and spending more time on topics like design, architecture, metrics, and team function. Consequently, the manager of the larger team may be more inclined to put real effort into the processes that make their team efficient.

Interactions *between* teams that don't necessarily work in the same way, but which contribute collaboration on the same project, highlight the importance of good communication skills and the role of data analysis and insights to drive agreement.

How Good Things Emerge

Let's get the semantics out of the way first. "Emerge" generally means that something comes into view that was previously occluded. This is not the nature of the process we are talking about here. Rather, we are capturing the idea that the form of an artifact (plan, design, architecture) is a product of multiple individuals interacting and iterating over time.[5] In the field of agile development, we are tapping into two key ideas: first, a self-organizing team can determine for itself the best subset of individuals to drive design and other conceptual work in specific areas as and when they become important; second, a design will improve through iterations of statement, criticism, and adjustment.

The right subset of individuals to drive a design varies with context and the relative importance of the component and state of work. Being able to establish if the design in question is critical (e.g., customer facing, relating to a component that will be hard to modify in the future, etc.) greatly informs the process. The most important designs should solicit information not just from within the team, but also from partners, customers, colleagues in other teams who have worked with the putative solution platforms, and so on.

Take, for example, a project involving determining the low-level serialization for a logical document representation. An initial implementation was created within the team while the system for inferring the logical document structure was being developed. This early implementation proved extremely useful for the development phase where inspecting the data and comparing it with the results of earlier models was a priority.

---

[5]This is closer to the artistic notion of emergence.

It also aligned with several existing tools and so made it easy to integrate into existing practices. However, when it came to delivering the data product to partners, it was clear that the performance of the representation in terms of deserialization and enumerating content was far more important than the qualities of the design that we had appreciated during the early development stages. A new, compact format was designed through discussion within the team, and this was ultimately the shipping vehicle for the data. The shipped design, which included an API to interact with the basic representation, was intentionally scoped to leave out capabilities that we anticipated would be useful, but for which the consumer had not expressed a need. We had plenty of discussion about ensuring that our general approach was open to extension should the need for those anticipated capabilities arise.

We often find that it is only through implementation that the best designs suggest themselves, and so exploratory implementations can often happen in the margins culminating in at least an initial design idea. Various engineering decisions such as the platform and programming languages used in the project will influence the details of any design, and so design through implementation can be an efficient way of not only exploring the design space but integrating the constraints and opportunities of those external factors in real time during the process.

Another key part of our experience has been the relationship between components in our ecosystem. If a new type of data challenges internal tooling critical to the workflows that the team carries out frequently, it is important that you consider evolving the tools rather than contorting the design to fit the established systems. This type of flexibility is in a sense the core principle of agility.

Nurturing a Self-Organizing Team

The autonomy of the team is not something that happens simply by decree. You can't just let a team wander off by itself and assert that it is now self-organizing. It requires a certain attitude from all members of the team as well as guidance from anyone in a management role within the team or anyone in a hierarchical leadership role outside the team. As a manager of a self-organizing team, you should pay close attention to the process by which you get things done and how things can be developed and improved to get the most from the spectrum of individuals you work with.

One of the hardest principles to teach a team that is newly operating in an agile way is the "we're all in this together" principle. Individuals may be used to a model where they are told specific tasks to do and they accomplish those tasks at high quality and on time. In Agile, the important deliverable is not a particular task but a team goal or deliverable. Individuals need to look up from their specific task and see the bigger picture and figure out how their contribution can help to achieve the larger goal. They need to be willing to jump in and work on other tasks to help the team achieve their goal, even tasks that aren't in their area of expertise or comfort zone. They need to fully engage their brain in planning meetings and keep the bigger picture in mind. They need to become more accountable for estimations and hitting their sprint goals.

Often a team undergoes an awkward period where they aren't quite fully engaging their brains and are still task and individual accomplishment oriented rather than thinking about the larger deliverables and how the team can accomplish them. During this period, it is important to let the team flail to some extent and avoid the temptation to jump in and micromanage every task the team does.

One natural way this happens is as the team follows the processes described in Chapter 4: Aligning with the Business, around "sprint goals," the team inevitably will not hit many of their goals for the first several sprints. It is important to be transparent about the fact that the team isn't hitting their goals and encourage the team to think harder about how to set team goals that are achievable in a sprint. This exercise will help the team to work harder at estimation, planning, and organizing themselves to start hitting a larger percentage of their sprint goals sprint after sprint.

Engineering Principles and Conceptual Integrity

The general approach we have described so far – that design comes from interactive and iterative processes in the team – may seem to conflict with the well-known alternate expressed in *Fred Brooks' The Mythical Man-Month*. In particular, the notion of a chief architect or chief programmer who is accountable for the design of the system seems to conflict with Agile development. As in all things agile, processes themselves are flexible, and so we prefer to think of a spectrum of team types. At one extreme is the top-down, chief architect model. At the other is the fully fluid, emergent model. The reality is that teams will always have internal or external oversight, and it is a matter of judgment for individuals in those positions (managers, tech leads, etc.) to determine when and how ego-driven design should occur. Some types of more complex systems may need to have much more oversight than others. Conceptual integrity is certainly important; and so, like many of the topics we've touched on in this book, it is another non-coding skill that everyone should at least be aware of and at best figure out how they contribute to.

One area where we do feel a little more control is required, and rewarded, is in the engineering principles adopted by the team and across teams in a multi-team product organization. We've covered many of these principles in Chapter 9: Technical Excellence.

Conclusion

In Chapter 11, "Self-Organizing" Teams, we've described the personalities that can be found in many teams working on data projects. We've highlighted the diversity in the team and the dynamics that ensue and how this contributes to the processes that establish, iterate, and improve on the designs and other declarative artifacts produced by the team.

In Chapter 12, "Tuning and Adjusting", we will discuss techniques for looking back and reflecting on past progress with an aim to become more effective.

Tuning and Adjusting

*At **regular intervals**, the team **reflects** on how to become more effective, then **tunes and adjusts** its behavior accordingly.*

—agilemanifesto.org/principles

Looking Back

We've talked about several mechanisms in this book for looking back and reflecting on past progress with an aim to becoming more effective as individuals, managers, and teams. To recap those mechanisms, they include **Retrospectives**, **Data Wallows**, **Quality Reviews**, **Live Site Reviews**, **Engineering Reviews**, and **Surveys.**

Retrospectives as discussed in Chapter 4: Aligning with the Business are a regular opportunity to reflect on what has gone well and what needs improvement on the team. We have used the process of having a board with three columns: "Good," "Bad," and "Meh." Meh is a great modern word that is used to express indifference or mild disappointment. This is a good middle ground category that will sometimes get feedback where people are reluctant to go so far as to label an issue as being a "Bad" one just yet. In the Retrospective meeting, we give everyone a pad of sticky notes and about 10 minutes to think of as many "Good," "Bad," and "Meh" observations about the sprint as possible. Team members then put their sticky notes on the board. Then someone on the team looks through all the sticky notes in a column and organizes similar sticky notes into groups. The team then discusses each sticky note grouping and brainstorms on ideas to fix the issues that are expressed in the "Meh" and "Bad" columns as well as how to keep the good things going that are expressed in the "Good" column. Specific user stories are devised that can be added in the next sprint to address the issues discussed in the Retrospective.

© Eric Carter, Matthew Hurst 2019
E. Carter and M. Hurst, *Agile Machine Learning*, https://doi.org/10.1007/978-1-4842-5107-2_12

Of course, a Retrospective is of limited use if no action is taken as a result, so it is important for the team to prioritize and complete action items determined as a result of Retrospectives. In addition to regularly completing user stories inspired by retrospectives, slack weeks as discussed in Chapter 8: Sustainable Development are another opportunity to work on user stories discovered through retrospectives.

In addition to considering the prior sprint in terms of how well specific stories or tasks were executed, we use retrospectives to address two other elements of the process. The first is interruption. By discussing work that was unplanned, but which for a variety of reasons the team was obliged to pick up, the team can reflect on the context within which they operate and any adjustments to planning or communication that could be made to better insulate the team from interruptions, thereby reducing uncertainty from the sprint. We have found cases of additional work cropping up where we have failed to do a good job of planning at the beginning of the sprint as well as where we have less than ideal automation and tooling support for certain parts of a live service.

The second is a core part of scrum – reflecting on the scrum process itself and any changes that the team can make to better execute. For example, if the team finds that mistakes were made due to the use of physical artifacts (sticky notes, a whiteboard used to capture state, etc.), they may decide to take the action of adopting a digital board system. Another example might be in adjusting the sprint length if the current length is determined to be too long or too short due to some aspects of the project having more or fewer unknowns. Team members might also discuss general performance impacting issues such as the number and frequency of meetings which they and their manager can immediately act on. A key attitude to this part of improvement is not to attempt to fix everything at once, but to make some clear, real changes. It's great for a team to see the process issues that they care about being continuously improved.

Data wallows as discussed in Chapter 6: Effective Communication are another great way to find opportunities to become more effective. A number of skills are used in these active, data-focused collaborations. Team members have the opportunity to practice and improve meeting and presentation preparation, data science skills which will be interrogated (in helpful and constructive ways) by the rest of the team, driving for clarity in determining and taking action based on data, the use of any number of applicable data presentation and manipulation tools, and, of course, the key skill of running an effective meeting.

Quality reviews as discussed in Chapter 2: Changing Requirements should be scheduled regularly so the group looks at and considers patterns of DSATs and what is being done to address those patterns. This is also an opportunity to look at trends in key metrics and make specific plans to continue improving those trends or address and fix any metrics that are trending downward. Review the information you are receiving from your monitoring as discussed in Chapter 7: Monitoring to ensure your product is behaving as expected.

In Chapter 8: Sustainable Development, we discussed the *Live Site Review*. This is an opportunity to further dive into live site issues and make sure they are understood and addressed. In this meeting, the team focuses on time to detect, time to engage, and time to mitigate and finds ways to drive these numbers down.

Engineering Reviews are discussed in Chapter 5: Motivated Individuals. This is an opportunity to monitor the key metrics around developer agility including inner loop and outer loop times for engineering. This is also a good opportunity to consider the results from a regular developer NSAT survey.

And in general, *Anonymous Surveys* are another important tool for determining how the team is doing. Sometimes individuals are hesitant to bring up an issue in one of the team settings described above, but they will respond to a quick anonymous survey about a particular issue or theme. Use surveys to make sure you give everyone a chance to give their opinion, even those who are reluctant to speak out in a group setting.

The Five Whys

One useful technique the team should use when determining why something went wrong is known as the "Five Whys" technique. This is a technique that was used at Toyota as they developed their manufacturing processes. The technique recursively questions the cause of a problem until an actionable reason is found. The insight here is that the first explanation of why something went wrong is too shallow and doesn't get at the true root cause.

For example, in Bing local data, we had the following issue:

Problem: The customer reports they can't reach the business by phone.

Five Whys:

1. Why can't they reach the business? On calling the phone number on the web site, the phone number is wrong.

2. Why is the phone number wrong? On finding the right phone number for the business, the business reports they recently changed their phone number.

3. Why do we not have the changed phone number? After examining our source data providers, we see the phone number in two of our providers, but it wasn't picked as the final phone number.

4. Why didn't we pick the correct phone number? On evaluating the model for selecting the number, we find that the model picked the right number, but a correction was in the system that overrode the model.

5. Why did the system override the model? The system should have expired out an out-of-date correction that was older than the new correct value provided by a trusted provider.

In this case by going deep with the Five Whys (or you might need even more than five), we were able to determine that the real root cause for the bad phone number was a correction system that should have been aware that an old correction is of lesser value than a new data update from a trusted data provider.

Tuning Metrics

In addition to "tuning behavior" as described in the Agile manifesto, it is important to continue to tune metrics as well. The team will naturally over time begin to get better and better on their metrics. Most metrics are at best an approximation of the user experience expressed as a number. The approximation can be a successful measure to optimize against, but as you get closer and closer to "perfection" on a metric, the approximation can begin to degrade. At some point, it is worth tuning the metric to make it more representative of the customer experience.

For example, we have mentioned in this book that one of the metrics we had in Bing's local data team was a composite metric called Q. This metric aggregated together several content metrics: the business name precision, phone number precision and coverage, address precision and coverage, and open precision. We calculated this metric on our data and on Google's data. Over time, our "Q" number began to approach Google's Q number. But from a user experience standpoint, the number did not reflect the actual experience on Bing vs. Google – although we were better in some ways, we often fell short. The "Q" metric wasn't appropriately representing the gap between Bing and Google.

We did a series of analyses to develop a new metric which we called Qv2. Qv2 incorporated a bunch of additional factors that we realized were important in measuring our user experience such as duplicate rate and precision and coverage of the latitude and longitude of the business. Qv2 also added more specific judging guidance to further tune the way we judged that a name, phone number, or address was correct. As an example, for a chain business like McDonald's, we no longer gave ourselves as much credit if we could only point to the root domain of McDonald's (`www.mcdonalds.com`) but reserved full credit for a URL being correct only if it pointed to the specific store page of the business in question (`www.mcdonalds.com/us/en-us/location/wa/bellevue/...`).

Looking Forward

In addition to having regular times to look backward and evaluate progress and areas for improvement, it is also useful to have regular times to look forward and set new goals and tighten up business objectives and metrics.

In our teams, we've found a quarterly "product team meeting" to be a useful forcing function to help us to regularly look forward and revise the plan for the team. This is a regular opportunity to revise

- Product vision (as described in Chapter 5: Motivated Individuals)

- Top-level metrics

- Business objectives and goals

- Engineering system objectives

- Process objectives

- Live site objectives

It is amazing how much these things can shift within a 3-month period on many teams. So it is valuable to always have a current "master plan" document for the team that summarizes these areas and is regularly updated each quarter and can be referenced by new team members or external stakeholders. This can be as simple as a wiki page, a PowerPoint deck, or whatever is convenient – what is important is to have some agreed-on way of documenting the latest product team master plan.

Conclusion

At this point, we have accrued a wealth of tools to be able to reflect on how to become more effective and drive improvements in an Agile data engineering team. We discussed tools like Retrospectives, Data Wallows, Quality Reviews, Live Site Reviews, Engineering Reviews, and Surveys. We also discussed the use of the Five Whys to properly root cause issues on the team. Metrics are only approximations of the user experience and must be updated regularly to better reflect the user experience over time. Finally, a regular team meeting to force the team to update the team's written master plan for vision, metrics, business objectives, engineering system objectives, process objectives, and live site objectives is an effective way to keep a clear focus on what the team needs to achieve next.

CHAPTER 13

Conclusion

What follows are the principles of the Agile Manifesto (agilemanifesto.org/principles) with our suggested modifications for Agile machine learning teams underlined. We also repeat the recap for each chapter.

1. Our highest priority is to **satisfy the customer** through **early** and **continuous** delivery of **valuable data**.

   In Chapter 1, "Early Delivery", we looked at ways in which a team can get up and running on a new inference project. We touched on the central role of metrics and a culture that is biased to action to enable quick iterations using data analysis to inform the evolution of all aspects of the project.

2. Welcome **changing requirements**, even **late in development**. Agile processes harness change for the customer's **competitive advantage**.

   In Chapter 2, "Changing Requirements", we discussed how you can build and develop systems anticipating change; tests, monitoring, and measurement to anticipate and measure change; and strategies for responding to DSATs and measured problems in your system.

3. *Deliver* **working software and accurate data continuously**, *from a couple of weeks to a couple of months, with a preference to the shorter timescale.*

   In Chapter 3, "Continuous Delivery", we described techniques to not just deliver working software frequently but to deliver working software and accurate data continuously. These include techniques like applying sufficient rigor at development time

© Eric Carter, Matthew Hurst 2019
E. Carter and M. Hurst, *Agile Machine Learning*, https://doi.org/10.1007/978-1-4842-5107-2_13

to ensure code and data changes are correct, using continuous integration systems to further verify those changes, deploying continuously to an ever-increasing number of users to test changes in production, and having systems to quickly roll back data and code changes if something goes wrong. We also talked about using techniques like flighting combined with telemetry and "win/loss" analysis to decide what to ship.

4. **Business people and developers** must work together **daily** throughout the project.

In Chapter 4, "Aligning with the Business", we discussed the importance of communicating with the business team. We described how important it is to have teams that are aligned as directly as possible to business metrics and business goals. We have shared some thoughts on how to work with the business to help them understand the limitations of machine learning. We also described how we do Scrum and how we involve and communicate to the business team through scrum meetings, scrum artifacts, and emails around our scrum cadence.

5. Build projects around **motivated individuals**. Give them the **environment** and **support** they need, and **trust** them to get the job done.

In Chapter 5, "Motivated Individuals", we discussed how you can build projects around motivated individuals. We discussed the importance of rewriting frequently. We described some leadership ideas around how to set a vision and metric targets while trusting motivated individuals to figure out the best way to achieve the vision and metrics. We discussed how to find and hire motivated individuals and how to retain and grow their careers once they join. We discussed the importance of having a productive development environment and measuring inner loop and outer loop times for common tasks as well as developer NSAT. Finally, we talked about how to leverage motivated individuals outside your team (which also will motivate people within your team).

6. The most **efficient** and **effective** method of conveying information to and within a development team is **face-to-face conversation.**

   In Chapter 6, "Effective Communication", we illustrated and contrasted the principle with a dive into many of the types of interactions that you will encounter as an agile data engineering team. The main message here is that all of this communication is a skill – just as writing code and going deep on specific machine learning methods. Like any skill, it can't be learned, maintained, or improved unless it is acknowledged and becomes an intentional part of your personal work and the team's culture.

7. **Monitoring of** working software and data provides the primary measure of progress.

   In Chapter 7, "Monitoring", we discussed how activity-based monitoring can be used to truly measure your working software. We discussed how an activity-based monitoring system works and the kinds of things that it logs and tracks. We talked about Azure Data Explorer which allows you to do sophisticated ad hoc queries of large amounts of logging data. We examined the various kinds of things that good monitoring can tell you: about whether your software is really working as expected, about why and when the software is malfunctioning, about what performance users are actually experiencing from your system, about whether the business goals are being met, about whether your customers' needs are being met, and about how data and machine-learned models can be monitored in production.

8. Agile processes promote **sustainable development**. The sponsors, developers, and users should be able to maintain a constant pace indefinitely, but varying the pace up and down occasionally can produce an even better result.

   In Chapter 8, "Sustainable Development", we talked about how to determine if you are working too fast or too slow and how to adjust the pace down and up. We talked about how to set goals and ensure that goals are both sufficiently ambitious and also achievable. We talked about the importance of engagement on a team to get the pace to the right level. We also talked about the importance of having some changes of pace built into development schedules with slack weeks to slow down the pace and hackathons to speed up the pace. We also talked about special issues around sustainable pace in managing live site issues and in working with multiple geographies.

9. **Continuous attention** to **technical excellence** and **good design enhances agility**.

In Chapter 9, "Technical Excellence", we motivated the investment in excellence in software engineering in general – *optimize for developer productivity* – and discussed the parallels in data projects – *maximize the value of data*. It might be fair to say that the most common confusion about agile engineering practices is the idea that agility involves cutting corners and avoiding tasks that don't appear to be on a direct line to the stated objective. But like any endeavor, efficiency of execution comes from a solid foundation in the basics. This is true of the foundational skills of the individual engineer, but also true of the foundations on which a team runs and the foundations on which a project operates. Early and continuous investment in quality will drive efficiency throughout the project.

10. Simplicity – the art of **maximizing the amount of work not done** – is essential.

In Chapter 10, "Simplicity", we described how diligence and focus in ensuring that the work being done clearly drives the project forward is the central goal of Simplicity. You are a member of, or a manager of, a team of smart data developers eager to make an impact on the world through your services and data products. You should come in to work every day with a keen eye for relevant work that is aimed at moving a project forward. You should challenge your customers to be engaged and present in the progress of the work. Within the team, you should help guide the progress through the unknown and un-navigated landscape of your data problem by helping to scope the problem, look for knowledge gaps, and articulate well-defined experiments.

11.   The best architectures, requirements, and designs emerge from self-organizing teams.

In Chapter 11, "Self-Organizing Teams", we described the personalities that can be found in many teams working on data projects. We highlighted the diversity in the team and the dynamics that ensue and how this contributes to the processes that establish, iterate, and improve on the designs and other declarative artifacts produced by the team.

12.   At regular intervals, the team reflects on how to become more effective, then tunes and adjusts its behavior accordingly.

In Chapter 12, "Adjusting and Tuning", we discussed tools like Retrospectives, Data Wallows, Quality Reviews, Live Site Reviews, Engineering Reviews, and Surveys. We also discussed the use of the Five Whys to properly root cause issues on the team. Metrics are only approximations of the user experience and must be updated regularly to better reflect the user experience over time. Finally, a regular team meeting to force the team to update the team's written master plan for vision, metrics, business objectives, engineering system objectives, process objectives, and live site objectives is an effective way to keep a clear focus on what the team needs to achieve next.

Index

A

Agile machine learning
    adjusting and tuning, 241
    business people and developers, 238
    changing requirements, 237
    continuous delivery, 237
    effective communication, 239
    monitoring, 239
    motivated individuals, 238
    principles of, 237
    self-organizing teams, 241
    simplicity, 240
    sustainable development, 239
    technical excellence, 240
Azure Data Explorer *vs.* database, 161

B

Business, 71
    advantages of colocation, 74–76
    driven scrum teams
        concrete business metrics, 78
        conversion rate, 77
        data providers, 80
        objectives, 76, 77
        understand data, 80
        web properties, 79
    English language version, 72
    importance of, 72–74
    machine learning, 81–84

    moving forward, 73
    physical collocation, 76
    scrum (*see* Scrum)
    strong competitor, 83
Business data
    analysis, 8–10
    attributes, 2
    distance learning approach, 7
    elements, 6
    establishment values, 10–12
    generic architecture, 5
    infrastructure, 5, 7
    investments, 4
    metric process, 2
    off-the-shelf approaches, 7
    representation of, 1
    schema, 2
    singleton businesses, 4
    value equilibrium, 12
    web mining components, 6

C

Change requirements
    building for
        architecture, 42–44
        correction layer, 30
        environments/rings, 29
        long-term change, 28
        measurement, 26–28

E. Carter and M. Hurst, *Agile Machine Learning*, https://doi.org/10.1007/978-1-4842-5107-2

D

E, F, G

Printed in the United States
By Bookmasters